Exercises in Multivariable and Vector Calculus

Exercises in Multivariable and Vector Calculus

Caspar R. Curjel
Professor of Mathematics
University of Washington, Seattle

Illustrations by James Flanagan

McGraw-Hill, Inc.
New York St. Louis San Francisco Auckland Bogotá
Caracas Lisbon London Madrid Mexico Milan
Montreal New Delhi Paris San Juan Singapore
Sydney Tokyo Toronto

WCB/McGraw-Hill
A Division of The McGraw-Hill Companies

This book was set in Computer Modern Roman by James Flanagan using $T_EXTURES$ on the Macintosh II computer.

The editors were Robert Weinstein and Margery Luhrs;
the production supervisor was Denise L. Puryear.
The cover was designed and produced by James Flanagan.
Semline, Inc., was printer and binder.

EXERCISES IN MULTIVARIABLE AND VECTOR CALCULUS

Copyright © 1990 by McGraw-Hill, Inc. All rights reserved. Printed in the United States of America. Except as permitted under the United States Copyright Act of 1976, no part of this publication may be reproduced or distributed in any form or by any means, or stored in a database or retrieval system, without the prior written permission of the publisher.

10 QPD/QPD 0 9

ISBN 0-07-014949-6

Library of Congress Catalog Card Number: 89-64436

Contents

Preface vii

The Mechanics of Using the Exercises 1

1 - Lines, Curves, Planes and Surfaces 3

Lines in the plane *3*; Lines in space *5*; Curves in the plane *7*; Curves in space *8*; Curves in non-cartesian coordinates *10*; Parameter representations of planes *11*; Parameter representations of surfaces *14*; Assembling parameter representations *15*.

2 - Functions of Two or More Variables 19

Linear functions of two and three variables *19*; Functions of two variables *20*; Functions of three variables *22*; Functions of n variables *22*.

3 - Partial Derivatives, the Chain Rule 25

Partial derivatives *25*; The chain rule *27*.

4 - Linearization 29

Gradient and directional derivative of linear functions *29*; Gradient and directional derivative of non-linear functions *31*; The linear approximation *34*; Numerical computations *36*.

5 - Optimization 39

Local and global extrema *40*; Reasoning from the definition of local extrema *43*; Functions of three variables *43*; Constrained extrema *44*; Setting up problems with Lagrange multipliers *46*.

6 - Polar, Cylindrical and Spherical Coordinates 49

Polar coordinates (r, θ) *49*; Cylindrical coordinates (r, θ, z) *51*; Spherical coordinates (ρ, θ, ϕ) *52*.

7 - Double Integrals 55

Riemann sums, estimates, mean value theorem *55*; Setting up double integrals in cartesian and polar coordinates *58*.

8 - Triple Integrals 63

Riemann sums, estimates *63*; Setting up triple integrals in cartesian coordinates *64*; Setting up triple integrals in cylindrical and spherical coordinates *66*.

9 - Vector Fields 67

Definition, components in cartesian coordinates *67*; The flow lines of a vector field *69*; The tangential component of a vector field along an oriented curve *70*; Components in non-cartesian coordinates *72*.

10 - Line Integrals 75

Line integrals of scalar functions - Riemann sums, estimates *75*; Line integrals of scalar functions - Evaluation *76*; Line integrals of vector functions - Riemann sums, estimates *77*; Line integrals of vector functions - Evaluation *78*; The various notations for $\int_C \vec{F} \cdot d\vec{r}$ *79*; Path independence *80*; The curl-test for vector fields in the plane *81*; Confusion surrounding the curl-test *82*.

11 - Flux and Circulation in the Plane, Green's Theorem 85

The normal component of a vector field along a curve in the plane *85*; Flux across a curve in the plane *86*; Green's theorem for flux and circulation in the plane *87*; Invariant definition of curl and divergence in the plane *90*.

12 - Surface Integrals, Flux Across a Surface 91

Surface integrals of scalar functions - Riemann sums, estimates *91*; Surface integrals of scalar functions - Evaluation *92*; The normal component of a vector field on a surface *93*; Flux of a vector field across a surface - Riemann sums, estimates *95*; Flux across a surface - Evaluation by inspection *97*; Flux across a surface - Evaluation by integration *98*.

13 - The Theorem of Gauss (Divergence Theorem) 101

The boundary of solids, surfaces and curves *101*; The theorem of Gauss (the Divergence Theorem) *102*.

14 - The Theorem of Stokes 109

15 - Gauss' and Stokes' Theorem, Miscellanea 115

Problems on both theorems *115*; Invariant definition of divergence and curl *116*.

16 - Coordinate Transformations 117

Linear transformations *117*; The Jacobian of linear transformations *121*; Curvilinear coordinates *122*; Change of variables in multiple integrals - Linear transformations *128*; Change of variables in multiple integrals - Nonlinear transformations *130*.

Answers 135

Ch. 1: *135*; Ch. 2: *142*; Ch. 3: *145*; Ch. 4: *147*; Ch. 5: *150*; Ch. 6: *153*; Ch. 7: *156*; Ch. 8: *159*; Ch. 9: *161*; Ch. 10: *166*; Ch. 11: *169*; Ch. 12: *171*; Ch. 13: *173*; Ch. 14: *175*: Ch. 15: *177*; Ch. 16: *178*.

Index 185

Preface

Description, general point of view. *Exercises in Multivariable and Vector Calculus* is a collection of exercises for the multivariable topics of a calculus course for the natural and engineering sciences. The exercises cover the following broad areas: Functions of two or more variables (linearization, optimization); double and triple integrals (including change of variables); and vector analysis (line and surface integrals, the theorems of Green, Gauss and Stokes). They are designed to be used, and have been successfully used, in conjunction with different standard textbooks, and they do not duplicate those textbooks' exercises.

With our exercises we address the following unsatisfactory state of affairs. Although most students with elementary calculus behind them quickly learn to manipulate formulas, only a few reach some understanding of the multivariable material in the sense of being able to work problems which are moderately different from problems they have seen before. At the end of the course, even many of the better students feel comfortable only with routine technical problems. I take this as a sign that in certain ways the teaching of multivariable calculus does not work properly, and the question arises what could be done about it.

One course of action is to focus on the problems to be worked by students. Roughly speaking, in a typical textbook there are two kinds of end-of-chapter exercises, viz. the "ordinary" exercises and the "difficult" ones. The ordinary exercises deal mostly with narrow parcels of formula work and are close to examples discussed in the text. The difficult exercises usually require considerable mathematical maturity and remain out of reach of most students. What is missing is a middle ground - exercises which presuppose fluency with formula work, but which train students to cope with moderately new situations. Exercises of this kind should help students develop skills at sorting out information, perceiving relations, separating issues and making decisions involving the notions and processes of multivariable calculus. The present collection provides a supply of such middle ground exercises.

Some features of our exercises. Given the specific agenda of our exercises, we found it necessary to make use of formats rarely found in exercises of published texts. Some of these less usual formats are described as follows.

The data may be given partially or fully in graphical form. Since there is little or no formula work required which would provide guidance, the student must go back to the meaning of the mathematical concepts themselves. Examples: (1) In **2.6** (p. 20) a traveling wave $u(x,t) = f(x-t)$ is discussed where the function f is given by a drawing. In the absence of formulas, questions such as "at which time t is the point $x = 73$ displaced by 1 unit for the last time?" have to be answered by looking at the way the disturbance travels. (2) In **5.9** (p. 44) a function $f(x,y)$ is given by some level curves. The question is to locate the critical points of $f(x,y)$ along several curves given by drawings. In answering the questions the student will work through the ideas behind the Lagrange multiplier. (3)

In **16.13** (p. 126) some coordinate curves of a system of curvilinear coordinates are given in a drawing, and questions are asked on the Jacobian determinant.

In general, a geometric-visual, if not tactile, approach is stressed. Many topics are geometric by their very nature, and most students have had little training in geometric reasoning. Examples: (1) In **4.10** (p. 33) the value of a function g at a point P and its gradient at the same point P are given. The question is to estimate the value of $g(Q)$ for points Q on various straight lines which pass through P. (2) Answering **12.10** (p. 95, estimate of a surface integral) requires looking at a vector field and asking questions like "what is the field vector at different points; how does it vary?" (3) In **13.1** (p. 101) the student is asked to describe the boundary of solids, surfaces and curves as preparation for the theorems of Gauss and Stokes.

As an alternative feature, a problem may consist of several interconnected parts which are to be worked in sequence. This makes it possible to ask broader and deeper questions. Examples: (1) **4.18** (p. 36) works through the various objects connected with a function of four variables: directional derivative, linear approximation, level set of the function and its linear approximation, parameter representation of the latter. – (2) In **11.10** (p. 88) the student has to work with Green's theorem in several related situations and ends up formulating a general fact about the vector field in question. – (3) In **15.1** (p. 115) several surfaces and curves are considered, all related to the same halfsphere. The problem is to compute the flux and circulation of a vector field and its *curl*. The student has to decide which of the transformation theorems to use and, above all, not to confuse them with each other.

Context of the problems, applications. The problems are formulated in mathematical terms. The only notions of physics we work with are "velocity vector of a moving point," "density," "mass," and "flux of a vector field across a surface." These notions are sufficiently close to the everyday material world so that students can be expected to develop an intuitive and operational understanding of them. Other uses of physics terms are only terminological (e.g., using "circulation" for "line integral over a closed curve").

We deliberately excluded problems involving more sophisticated physical notions such as moment or charge density for two reasons. First, we think that having students work with notions which in a mathematics course can be handled at the formal level at most will only reinforce the myth "Mathematics is mostly formulas, possibly very complicated ones." Second, our exercises can be worked by students irrespective of their past or future areas of specialization.

From the point of view of actual applications of multivariable calculus in other fields several questions such as **15.3** (p. 116) on the invariant definition of divergence may look unrealistic since they would not arise in real situations. However, these questions are likely to help students develop some degree of conceptual familiarity with, or feeling for, the notions under discussion, and they reflect the flavor of authentic applications in the concrete way the various ingredients of the situation are articulated.

Classroom use of the exercises. Our exercises are meant to be used in conjunction with a published text which provides (i) exposition of the theory, and (ii) the usual technique-oriented exercises called ordinary above. These latter exercises are a prerequisite for our exercises.

The exercises can be used in an existing course without undue effort because their organization parallels the organization of many published texts. For a period of two years we have used most of our exercises with several texts from various publishers, covering the full-length calculus sequence as well as vector calculus and advanced calculus. Students worked the problems in work periods during lectures in the presence of the author, as homework, or in examinations.

The exercises worked well in the classroom. They require mathematical reasoning at a realistic level and definitely prohibit a view of mathematics as a set of rules to shuffle anonymous symbols. We found that the problems are within reach of students who make a solid effort in the course; that students seem to derive satisfaction from working them; that they generate productive discussions among students and lend themselves well to work in groups; and that the questions are mostly free from extraneous complications (such as excessively complicated formula work) which detract from the main thread of ideas in an unproductive way.

Given the unusual mode of some of the exercises, students and fellow instructors have asked questions–as will future users–concerning their administration. In the section "The Mechanics of Using the Exercises" we discuss issues typically raised by students and instructors alike.

Last but not least, the problems seem to work for the instructor, too. For example, I find it rewarding to observe a group of non-math majors working on a problem like **4.10** (p. 33) and discussing what a directional derivative is, why and how to use it, and how the gradient enters. From a more practical point of view, the questions of many problems can be easily varied in a meaningful way so that the pool of practicable problems for lectures and examinations is greatly enlarged.

Acknowledgments. The present collection is an outgrowth of work done jointly with G. S. Monk several years ago. Many of the ideas in the exercises are either offsprings of ideas of his, or owe their existence to his unique gift of helping others crystallize their insights. I would like to thank him for all I learned from him. Also I would like to thank Heinrich Matzinger for the many suggestions and probing questions he brought up while we worked together for a year. Several parts of the present collection began taking shape in the course of our joint work.

I wish to thank the reviewers Richard Bagby (New Mexico State University), Elizabeth Appelbaum (Webster University), Theodore Faticoni (Fordham University), Piotr Mikusiński (University of Central Florida), and Jim Hefferon (Union College) for their advice, and acknowledge the help of Ward A. Sooper (Walla Walla College) and Eldon L. Miller (The University of Mississipi) who checked the answers.

Last but not least, the exercises would not have reached the stage of publication if it had not been for Mitch Beaton and Robert Weinstein, both at McGraw-Hill, and Jim Flanagan, a former student of mine. Mitch Beaton initiated my contacts with McGraw-Hill and was always available when help was needed. Robert Weinstein was my Sponsoring

Editor. I greatly appreciated his support, his patience and, above all, his exemplary way of mediating between the worlds of academia and business. Jim Flanagan did the art work and the design. His rare combination of skills and interest in mathematics, computers and graphic design proved invaluable for the project, and working with him was enjoyable as well as stimulating. It is a special pleasure to thank Messrs. Beaton, Weinstein and Flanagan for their efforts on my behalf.

Caspar R. Curjel

Exercises in Multivariable and Vector Calculus

The Mechanics of Using the Exercises

Notation of points and vectors. Unless otherwise mentioned we work in a cartesian xyz-system. A point P with coordinates $(x,y,z) = (a,b,c)$ is written as $P(a,b,c)$.

As basis for vectors we take $\vec{\imath}$, $\vec{\jmath}$ and $\vec{k}$ and write $\overrightarrow{OP} = (a,b,c)$ for $\overrightarrow{OP} = a\vec{\imath} + b\vec{\jmath} + c\vec{k}$.

Exact answers. Exercises whose data are given in graphical form are meant to be worked by ruler and pencil as well as circumstances permit. At first, many students are bothered by such uncertainty. If they are told right away

- that they are expected to draw lines and read graphs, distances, etc. as well as they can, and
- that in view of variations in eyesight and dexterity with drafting utensils there will always be a *range* of correct answers,

they adapt quickly.

In Chapter 17 the answers to ruler and pencil problems are given in the form of equalities. For example, if by measurements (or by a combination of measurements and computations) t was found to be 70.8, we write $t = 70.8$ (and not $t \approx 70.8$).

Exact hypotheses. In the exercises we do not state the exact hypotheses which the objects under discussion have to satisfy. The functions, curves, etc. are always tacitly assumed to be sufficiently regular for the exercise to proceed as intended.

"None," "Impossible," etc. as the answer. There are questions which have infinitely many correct (and infinitely many incorrect) answers. In this situation students have to pick an answer. For the first few times they find it difficult and need the instructor's encouragement. Then there are questions for which "None" or "Impossible" is the answer. Questions of this kind are often perceived as trick questions. They are not. In such a problem the relation between the question and the data is different from the relation in a problem which admits a solution. Analyzing a set-up and concluding that it is contradictory is a mathematical task as legitimate as working a problem which has one or more solutions. — Students should be told to expect problems with many solutions and others with no solution.

The use of isometric axonometry. All 3-D illustrations are done in isometric axonometry (orthogonal parallel projection with ray of vision parallel to $\vec{a} = (-1,-1,-1)$), as shown in Fig. 1 on p. 2.

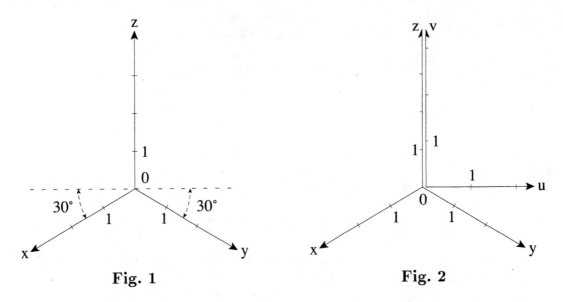

Fig. 1 **Fig. 2**

The projections of the units on the three axes are all of the same length. This makes isometric axonometry suitable for use with students who have little experience in 3-D sketching because it is easy to draw points, lines etc. in a realistic, i.e., correct way. The projections of circles in the coordinate planes are isometric ellipses (axes of ratio $\sqrt{3}:1$). In order to produce projections of circles in general position and of other objects it is convenient to use the system of uv-coordinates in the plane of projection shown in Fig. 2 above right. A point $\overline{P}$ in space with coordinates $(\overline{x},\overline{y},\overline{z})$ projects onto a point P whose uv-coordinates are given by

$$u = -\frac{1}{\sqrt{2}}\,\overline{x} + \frac{1}{\sqrt{2}}\,\overline{y},$$
$$v = -\frac{1}{\sqrt{6}}\,\overline{x} - \frac{1}{\sqrt{6}}\,\overline{y} + \frac{2}{\sqrt{6}}\,\overline{z}.$$

Note that the units on the uv-axes are the same as the $\overline{xyz}$-units in space. They appear bigger than the projected xyz-units (ratio $\sqrt{1.5}:1$).

1 | Lines, Curves, Planes and Surfaces

Terminology – Note to instructor. The problems stress parametric equations. We simply call them "parameter representations," as in "Find a parameter representation of the plane defined by …." While this terminology may be new to some, it is concise and suggestive of "parameter" being a notion of its own, pointing at the distinction between a point set and its parametrizations. – "Line" always means "straight line."

Lines in the plane

■ **1.1.** Find a parameter representation $x = a + bt$, $y = c + dt$ of the following lines m_1, m_2, m_3:

m_1 is the line shown in Fig. 1.1 below;

m_2 passes through $A(1,3)$ and $B(4,0)$;

m_3 passes through $A(1,3)$ and is perpendicular to the line $5x - 3y + 2 = 0$.

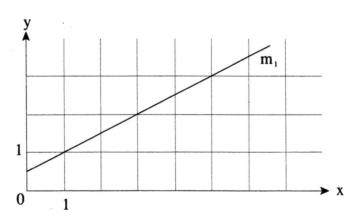

Fig. 1.1

■ **1.2.** The line m is given by the parameter representation

$$x = 1 + 3t, \ y = -2t.$$

a) Draw m into an xy-system and mark on m the points which correspond to $t = -2, -1, 0, 1, 2$.

b) Find the distance between the following two points R and S on m:

$$R: t = 896,312,464.03, \ S: t = 896,312,467.29.$$

c) Find the equation of m in the form $Ax + By + C = 0$.

1—Lines, Curves, Planes and Surfaces

■ **1.3.** The line k is given by
$$x = c_1 + \frac{3}{5}t, \quad y = c_2 - \frac{4}{5}t.$$

a) The point P_1 with coordinates (c_1, c_2) lies on k. Find all points P on k which are at a distance 15 from P_1.

b) Do **1.2** b) for k.

■ **1.4. a)** Find a parameter representation of the line $5x - 2y + 6 = 0$.

b) Use x as parameter for a parameter representation of the line $y = -2x + 4$.

c) Use y as parameter for a parameter representation of the line of b).

■ **1.5.** A point P moves on the line m shown in Fig. 1.5. We write $P(t)$ for its position at time t seconds. The position vector of $P(t)$ is given by
$$\overrightarrow{OP}(t) = \overrightarrow{OA} + t\overrightarrow{AB}.$$

a) Mark on the line the points $P(0)$, $P(1)$, $P(\frac{4}{5})$.

b) At which time t is $P(t) = S$?

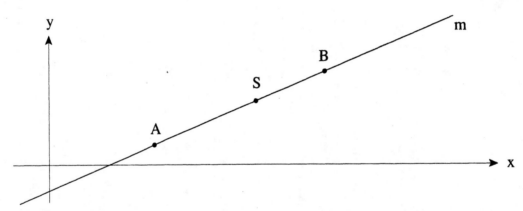

Fig. 1.5

■ **1.6.** We continue working with the line m of the preceding problem **1.5**. First, we note that the expression for $\overrightarrow{OP}(t)$ is nothing but a parameter representation of m. Second, we write R for $P(\frac{4}{5})$ and use R together with the point S of **1.5** b) for a second parameter representation of m:
$$\overrightarrow{OP} = \overrightarrow{OR} + u\overrightarrow{RS}$$
(we write u for the parameter to avoid confusing it with t). Now one and the same point P on m has a "t-coordinate" and a "u-coordinate":
$$\overrightarrow{OP} = \overrightarrow{OA} + t\overrightarrow{AB} = \overrightarrow{OR} + u\overrightarrow{RS}.$$

a) (i) Find a scalar c so that $\overrightarrow{RS} = c\overrightarrow{AB}$. (ii) Express $\overrightarrow{OR}$ in terms of $\overrightarrow{OA}$ and $\overrightarrow{AB}$.

b) Use a) to express the right side of $\overrightarrow{OP} = \overrightarrow{OR} + u\overrightarrow{RS}$ in terms of $\overrightarrow{OA}$ and $\overrightarrow{AB}$. Compare your result with $\overrightarrow{OP} = \overrightarrow{OA} + t\overrightarrow{AB}$ to obtain a formula which relates the t-coordinate t of P with its u-coordinate u.

■ **1.7.** A point P moves along a line g shown in Fig. 1.7; we write $P(t)$ for the position of P at time t seconds. The direction of movement and $P(0)$, the position of P at time $t = 0\,sec$ are as shown. We are told that $P(t)$ is at distance $t^3/5\,cm$ from $P(0)$.

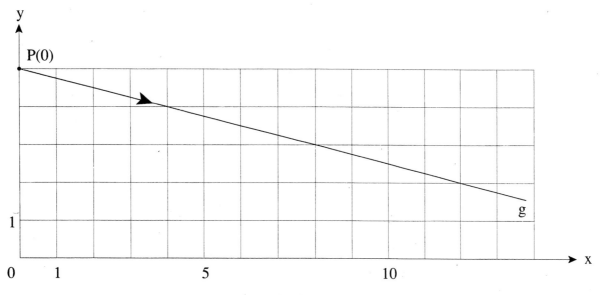

Fig. 1.7

a) Draw $P(t)$ for $t = 1, 2, 3, 4\,sec$.

b) Use a) to decide: Between $t = 0$ and $t = 4$ the speed of P is constant – increasing – decreasing – none of the these three.

c) For a moment go back to the point P moving on the line m of **1.5.**: Is P moving on m at constant speed? Increasing speed? Decreasing speed?

d) We continue with P moving along g of the present problem. The coordinates of $P(t)$ are $(x(t), y(t))$. Find the functions $x(t)$, $y(t)$.

Lines in space

■ **1.8.** The line m in space is given by the parameter representation

$$x = 2 + (t/4), \quad y = 3 - (t/2), \quad z = 1 + (3t/4).$$

a) Draw m into an xyz-system and mark on it the points which correspond to the parameter values $t = -1, 0, ..., 5$.

b) Decide whether or not the point $P(27, -47, 75)$ is on m.

■ **1.9.** The line k passes through $A(1, 0, 10)$ and $B(-3, -2, 20)$. M is the midpoint of AB, and D is the point on k such that $AD = 3 \cdot AB$ as shown in Fig. 1.9 (the coordinates of the points $A, B, \ldots$ are not to scale).

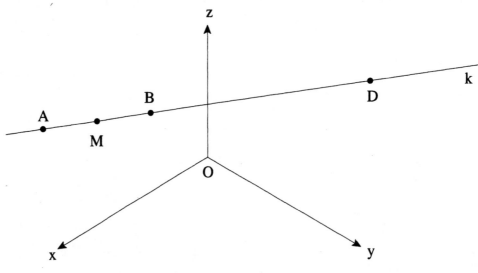

Fig. 1.9

a) Write down a parameter representation of k, using t as the letter for the parameter.

b) Now we consider the parameter values t as points on a t-axis which is separate from the xyz-space in which the line k lies. Identify the points on the t-axis to which the stretch MD on k corresponds.

c) What is the length of the segment MD in space? What is the length of the corresponding stretch on the t-axis which you found in b)?

■ **1.10.** Find a parameter representation $x(t) = a + bt$, $y(t) = c + dt$, $z(t) = e + ft$ of the line k with the following properties:

$$k \text{ is parallel to } \vec{m} = (5, 4, 2); \ (x(2), y(2), z(2)) = (3, 5, 0).$$

■ **1.11.** A point P moves at constant speed along the line h which passes through $A(4, 4, 5)$ and $B(1, 0, 2)$. The position of P at time $t\,sec$ is denoted by $P(t)$. We have the following information on $P(t)$: $P(0) = A$ and $P(1) = B$.

a) Find the coordinates $(x(t), y(t), z(t))$ of $P(t)$.

b) What is the speed of P on h (measured in units of length per second)?

c) The distance between $P(0)$ and $P(t)$ is a function $s(t)$ of t (this function is called "arc length on h"). Find $s(t)$.

d) At some time t^* the point P will cross the xy-plane. Find t^* and the position of P at time t^*.

e) Find another parameter representation of h with parameter u:

$$x(u) = a + bu, \ y(u) = c + du, \ z(u) = e + fu \ (a, \ldots, f \ constant),$$

such that the distance between $(x(u), y(u), z(u))$ and $(x(\overline{u}), y(\overline{u}), z(\overline{u}))$ equals $|\overline{u} - u|$.

■ **1.12.** The line m passes through $A(3, 1, 5)$ and is parallel to $\vec{b} = (1, 2, 1)$. Find the points in which m intersects the cylinder $x^2 + z^2 = 4$.

Curves in the plane

■ **1.13.** We consider three curves C in the plane and a point P on each of them:

$C_1: y = 3x^3 - 2x^2$, $P(2, 16)$;

$C_2: x = t^2 - t + 4$, $y = 3t^3 + 2t$, $t = 2$;

$C_3: 4x^2y - 3xy + x^2y^2 = 6$, $P(1, 2)$.

Find a parameter representation of (i) the tangent to C at P; (ii) the line normal to C at P.

■ **1.14.** C is the curve shown in Fig. 1.14. We denote by s the arc length on C counted from $(0, 2)$ on. Then the coordinates (x, y) of a point on C are functions of s: $x = a(s)$, $y = b(s)$. In other words: The functions $a(s)$, $b(s)$ provide a parameter representation of C.

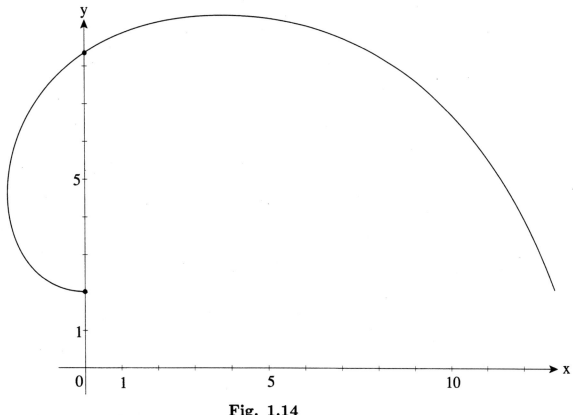

Fig. 1.14

a) Sketch the graphs $w = a(s)$, $w = b(s)$ into an sw-system (units 1 cm). Hint: To measure the arc length take a piece of paper with a straight edge and roll it along the curve C; allow 24 cm for the s-axis.

b) Draw the vector with components $(a'(6), b'(6))$ into an xy-system. Measure its length. What should the length be according to theory?

■ **1.15.** C is the curve $y = \cosh x$, considered only for $x \geq 0$.

   a) Find a parameter representation $x = p(t)$, $y = q(t)$ of C.

   b) Write $s(t)$ for the arc length on C counted from $(0, 1)$ on. Find $s(t)$.

   c) Use b) to find a parameter representation $x = a(s)$, $y = b(s)$ where s is the arc length on C counted from $(0, 1)$ on (remember that we consider C only for $x \geq 0$).

Curves in space

■ **1.16.** A point $P(x, y, z)$ moves on a curve C in space:

$$x = a(t),\ y = b(t),\ z = c(t)\ (t : time),$$

and the graphs of $a(t)$, $b(t)$, $c(t)$ are given in Fig. 1.16.

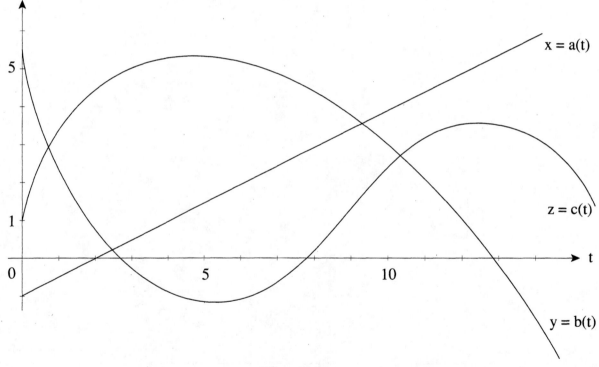

Fig. 1.16

Answer the following questions in the sense of "giving the best possible estimates on the basis of the information provided."

   a) Find the coordinates of P at time $t = 10$.

   b) Find the velocity vector and the speed of P at time $t = 2$.

   c) Use a tangent vector to estimate how many units P travels on C between $t = 2$ and $t = 2.1$.

■ **1.17.** C is a circle of radius 3 in the plane $y = 1$ with center at $(0, 1, 0)$, as shown in Fig. 1.17.

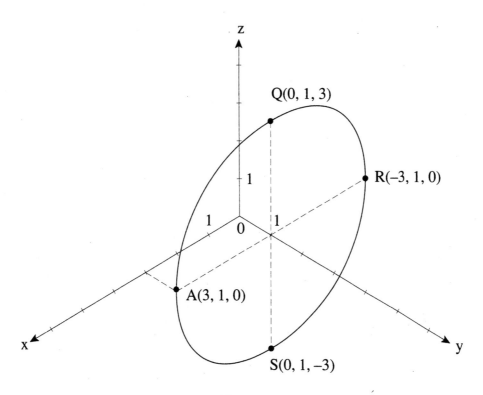

Fig. 1.17

a) Write down a parameter representation $(x(t), y(t), z(t))$ of C.

b) Now we consider the parameter values t as points on a t-axis which is separate from the xyz-plane in which C lies. Identify on the t-axis the points to which the halfcircle QRS corresponds, with Q, R, S as shown.

■ **1.18.** C is the curve given by $x = \frac{1}{2}t^2 + 3$, $y = \frac{4}{3}t^{3/2}$, $z = 2t + 5$. Throughout this problem we take $t = 1$ and $\Delta t = 0.03$. Note that Fig. 1.18 is not to scale at all.

a) Use $|\Delta \vec{r}|$ to give an approximation for the arc length from t to $t + \Delta t$.

b) Use $|\vec{r}\,'(t)\,\Delta t|$ for the same task.

c) Find the arc length by integration and compare with a) and b).

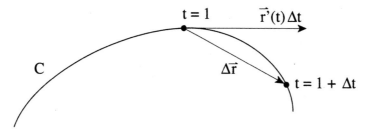

Fig. 1.18

■ **1.19.** C is the curve $x = 4t^3 + t^2 + 3$, $y = 2t + 1$, $z = -t^3 + 2$, and P is the point corresponding to $t = 1$. Find: (i) The coordinates of P; (ii) the point where C intersects the xz-plane; (iii) a parameter representation of the tangent line at P; (iv) the equation of the plane normal to C at P in the form $Ax + By + Cz = D$.

Curves in non-cartesian coordinates

We use the following notation and conventions for polar, cylindrical and spherical coordinates:

Polar coordinates are written (r, θ), i.e., $x = r\cos\theta$, $y = r\sin\theta$, $r \geq 0$;

Cylindrical coordinates are written (r, θ, z), i.e., $x = r\cos\theta$, $y = r\sin\theta$, $z = z$, $0 \leq \theta < 2\pi$, $r \geq 0$;

Spherical coordinates are written (ρ, θ, ϕ), i.e., $x = \rho\cos\theta\sin\phi$, $y = \rho\sin\theta\sin\phi$, $z = \rho\cos\phi$, $\rho \geq 0$, $0 \leq \theta < 2\pi$, $0 \leq \phi < \pi$.

Note that some authors interchange the letters θ and ϕ in spherical coordinates.

■ **1.20.** The curve C is given in polar coordinates (r, θ) by

$$r = 2\theta + \frac{\pi}{2}, \quad 0 \leq \theta \leq \frac{3\pi}{2},$$

i.e., C is part of a spiral with end points $(x, y) = (\frac{\pi}{2}, 0)$ and $(x, y) = (0, -\frac{7\pi}{2})$.

   a) Find a parameter representation in cartesian coordinates $x = a(t), y = b(t)$.

   b) Set up the integral which gives the arc length of C.

■ **1.21.** The curve C is given in cylindrical coordinates (r, θ, z) by

$$\theta = \pi/4, \quad z = f(r),$$

where $f(\ )$ is a function of one variable. C lies in the half-plane $x = y$, $x \geq 0$.

   a) Write down a parameter representation of C in cartesian coordinates.

   b) Write down a parameter representation (in cartesian coordinates) of the tangent line to C at the point $r = 2$.

■ **1.22.** The curve C is given in spherical coordinates (ρ, θ, ϕ) by

$$\rho = \sqrt{2}\,\theta, \quad (\theta \geq 0), \quad \phi = \pi/4,$$

i.e., C lies on the circular cone $z = \sqrt{x^2 + y^2}$.

   a) Find a parameter representation of C in cartesian coordinates.

   b) α is the plane normal to C at the point where $\theta = 2\pi$. Find the equation of α in the form $Ax + By + Cz + D = 0$.

Parameter representations of planes

■ **1.23.** The points $A(1,0,2)$, $B(0,2,2)$, $C(0,0,3)$ determine a plane α. As a parameter representation of α we take
$$\vec{r} = \overrightarrow{OA} + u\,\overrightarrow{AB} + v\,\overrightarrow{AC}.$$

a) Draw into an xyz-system: (i) the points A, B, C; (ii) the coordinate lines $u = -1, 0, 1, 2$ and $v = -2, -1, 0, 1$.

b) D is the point in α for which $(u,v) = (1,1)$. The points A, B, C, D determine a parallelogram which lies in α. Find its surface area.

c) Find the equation of α in the form $ax + by + cz + d = 0$.

■ **1.24.** Use x and y as parameters for a parameter representation of the plane $2x + 3y - 3z = 0$.

■ **1.25.** The plane β is given by the parameter representation
$$x = 3 + u + 2v,\ y = 1 - u,\ z = 2 + 3u + v.$$

a) Find a vector $\vec{b}$ which is normal to β.

b) Find a parameter representation of the line m in which β intersects the plane $x - y = 0$.

c) Decide which of these points lie in β: $(1, 0, 12)$; $(7, 1, 4)$; $(69, 15, 0)$.

■ **1.26.** T is the triangle shown in Fig. 1.26.

a) Write down a parameter representation $\vec{r} = \vec{a} + u\vec{b} + v\vec{c}$ of the plane β which contains T.

b) Now we consider the pairs (u, v) of parameters as points in a uv-plane which is separate from the xyz-space in which the plane β of T lies. Identify the region in the uv-plane to which the triangle T corresponds.

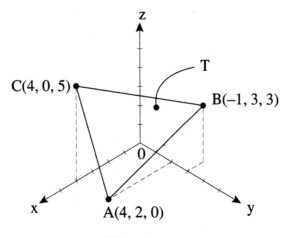

Fig. 1.26

■ **1.27.** The points A, B, C shown in Fig. 1.27 define a plane α. We take the following parameter representation of α:

$$\overrightarrow{OP} = \overrightarrow{OA} + u\overrightarrow{AB} + v\overrightarrow{AC}$$

(Note that the numerical values of the xyz-coordinates of A, B, C cannot be determined from the data.) Draw into the xyz-system below the point Q of α for which $(u,v) = (2, -1.5)$.

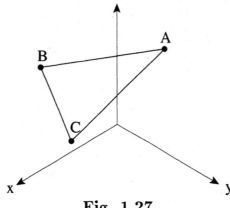

Fig. 1.27

■ **1.28.** We use the plane α of **1.23** together with the parameter representation given there.

E is the point of α for which $(u, v) = (2, 1)$. We use E together with A and C of **1.23** for a second parameter representation of α: $\overrightarrow{OP} = \overrightarrow{OA} + p\overrightarrow{AC} + q\overrightarrow{AE}$. Now one and the same point P in α has (u, v)-coordinates and (p, q)-coordinates:

$$\overrightarrow{OP} = \overrightarrow{OA} + u\overrightarrow{AB} + v\overrightarrow{AC} = \overrightarrow{OA} + p\overrightarrow{AC} + q\overrightarrow{AE}.$$

a) K is the point of α for which $(u, v) = (2, 0)$. Find the (p, q)-coordinates of K. Hint: Draw E and K to scale into Fig. 1.28 below. That should help you to see the answer.

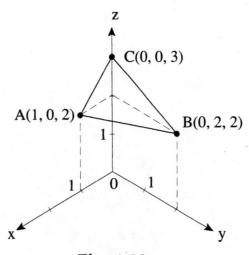

Fig. 1.28

b) Find the (u,v)- and (p,q)-coordinates of C.

c) Now we ask the following questions for a point P in α:

(i) Given (u,v) of P, find (p,q);   (ii) given (p,q) of P, find (u,v).

The statement "$(u,v) = (2,1)$ for E" means $\overrightarrow{AE} = 2\overrightarrow{AB} + \overrightarrow{AC}$. We use this expression for $\overrightarrow{AE}$ in the (p,q)-representation of α:

$$\overrightarrow{OP} = \overrightarrow{OA} + p\overrightarrow{AC} + q(2\overrightarrow{AB} + \overrightarrow{AC}).$$

At the same time the formula of the (u,v)-representation says:

$$\overrightarrow{OP} = \overrightarrow{OA} + u\overrightarrow{AB} + v\overrightarrow{AC}.$$

Compare the coefficients of $\overrightarrow{AB}$ and $\overrightarrow{AC}$ in these two expressions for $\overrightarrow{OP}$. You obtain two equations which involve u,v,p,q. Write these equations in such a form that they answer question (i). Then do the same for (ii).

d) Use the formulas you just obtained to confirm your answers to a) and b).

■ **1.29.** The two lines $h : (x,y,z) = (2,3,1) + t(4,2,1)$ and $k : (x,y,z) = (2,3,1) + t(1,1,2)$ determine a plane α (the sketch in Fig. 1.29 is not to scale). Find a parameter representation $\overrightarrow{OP} = \vec{a} + u\vec{b} + v\vec{c}$ of α such that h is the coordinate line $u = -1$ and k is the coordinate line $v = -1$.

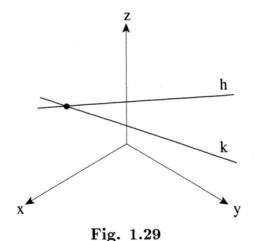

Fig. 1.29

■ **1.30.** α is the plane $x + y + 4z = 12$, and β is the plane given by

$$x = 5 + 2u + v,$$
$$y = 4 + 2v,$$
$$z = 10 + 3u + 5v.$$

Find a parameter representation of the straight line in which the two planes intersect.

Parameter representations of surfaces

■ **1.31.** S is the surface given by $x = v$, $y = u + v^2$, $z = u^2$. Several of the coordinate curves $u = const$ and $v = const$ of S are shown below right (Fig. 1.31 is *not* to scale).

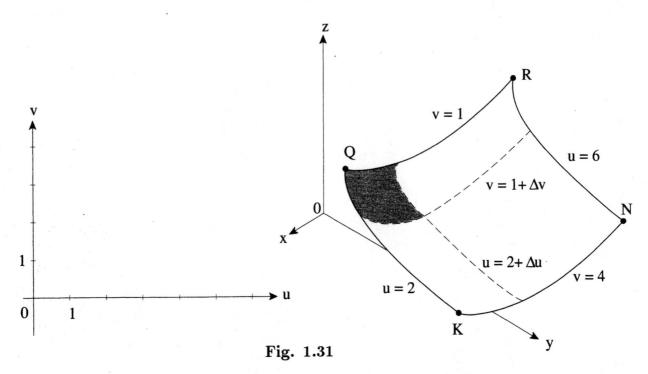

Fig. 1.31

a) Into the uv-plane on the left (the "parameter plane") draw the following: (i) The points Q_0, R_0, N_0, K_0 to which correspond the points Q, R, N, K on S; (ii) the lines to which correspond the coordinate curves $u = 2$, $2 + \Delta u$, 6 and $v = 1$, $1 + \Delta v$, 4 of S.

b) Give a parameter representation of the tangent line to the coordinate curve $u = 2$ at the point K.

c) Find a parameter representation of the tangent plane of S at the point K.

d) C_0 is the line in the parameter plane which passes through the points Q_0 and N_0. There is a curve C on S which corresponds to the line C_0 in the parameter plane. Give a parameter representation of C.

e) Set up the integral for the arc length of C from Q to N (please do not evaluate the integral).

f) Identify in the parameter plane the region to which the small shaded patch on S corresponds.

g) Use tangent vectors to the coordinate curves through Q to estimate (without integrating) the surface area of the shaded patch on S. Your answer will contain Δu and Δv.

■ **1.32.** S is the surface $z = xy$.

a) Write down a parameter representation of S with parameters x and y.

b) Find a parameter representation of the coordinate curves $x = 3$ and $y = 5$. What is the shape of these curves which lie on S?

c) You just realized that S contains many straight lines (i.e., there are many straight lines *all* points of which lie on S). Give an example of a surface S_1 which does not contain any straight line at all. Give an example of a surface S_2 which is not a plane but which contains many straight lines.

■ **1.33.** We consider three surfaces S and a point P on each of them:

$S_1: z = \frac{25}{x^2+y^2}$, $P(1,2,5)$;

$S_2: 2xy^2z + x^2yz^2 - z^3 = 2$, $P(1,1,1)$;

$S_3: x = (u-1)\cos v, y = (u-1)\sin v, z = v+1$, $P(2,0,1)$.

For each S find (i) the equation of the tangent plane at P in the form $Ax+By+Cz+D = 0$; (ii) a parameter representation of the normal to S at P.

Assembling parameter representations

■ **1.34.** The surface S looks like a spiral staircase (without the steps). It is generated as follows. At time $t = 0$ the ray $x \geq 0$, $y = z = 0$ starts turning around the z-axis while rising vertically at the same time. At time t the ray has turned by the angle $\theta = 2t$ and has risen $6t$ units.

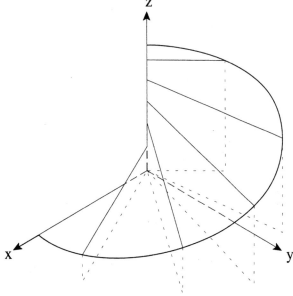

Fig. 1.34

a) Find a parameter representation of S with the r and θ of cylindrical coordinates as parameters (for cylindrical coordinates see the remarks preceding **1.20**):

$$x = f(r,\theta), \ y = g(r,\theta), \ z = h(r,\theta).$$

b) Give a parameter representation of the coordinate curve $\theta = \pi/3$. What is its shape?

c) Let P be the point on the curve $\theta = \pi/3$ which is 2 units away from the z-axis. Find the tangent plane to S at P.

■ **1.35.** C is a curve in the xz-plane. It is given by $z = f(x)$ and lies in the half $x \geq 0$ of the xz-plane. By rotating C around the z-axis we generate a surface S.

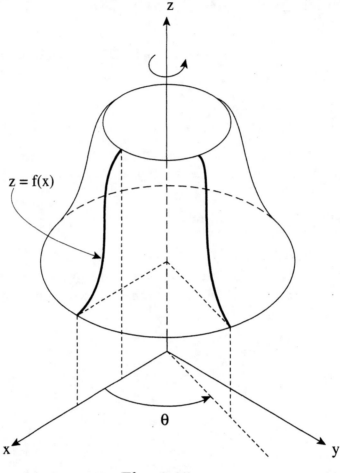

Fig. 1.35

a) Find a parameter representation of S with the r and θ of cylindrical coordinates as parameters:

$$x = f(r, \theta), \ y = g(r, \theta), \ z = h(r, \theta).$$

b) Describe in words the shape of the coordinate curves $r = $ *constant* and $\theta = $ *constant*.

c) Let P be a point on S whose location on S is characterized by the pair (r, θ) of parameter values. We consider a vector $\vec{n}(P)$ normal to S at P: $\vec{n}(P) = n_1\vec{i} + n_2\vec{j} + n_3\vec{k}$. The components n_1, n_2, n_3 are all functions of r and θ. Find these functions. – Note that this question has more than one correct answer. For if $\vec{n}$ is normal to S, then so are $-\vec{n}$, $16.73 \cdot \vec{n}$, etc. In other words, at a point P the normal *line* is uniquely defined, but the normal *vector* is not uniquely defined.

■ **1.36.** S is the straight circular cone shown in Fig. 1.36. We consider S as extending infinitely in the negative z-direction.

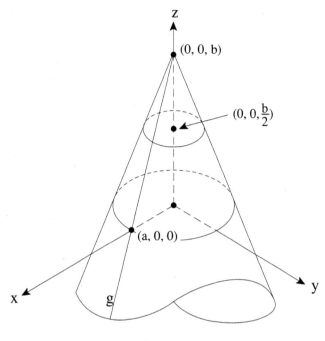

Fig. 1.36

a) Use the method of **1.35** to find a parameter representation $x = f(r, \theta)$, $y = g(r, \theta)$, $z = h(r, \theta)$ of S.

b) D is the part of S between the horizontal planes $z = 0$ and $z = \frac{b}{2}$. In the $r\theta$-plane (the "parameter plane") there is a region D_0 to which corresponds D. Draw D_0.

c) Find the equation of the tangent plane at $(a, 0, 0)$ in the form $Kx + Ly + Mz = N$ as follows: Look at the picture of the cone and reason from scratch, i.e., do *not* use the formula for normals worked out in **1.35 c)**.

d) Find a vector normal to S at $(a, 0, 0)$ via the normals formula of **1.35 c)**.

e) P is the point in which the line g (the line connecting $(a, 0, 0)$ and $(0, 0, b)$) intersects the horizontal plane $z = -6922.591$. P lies on the cone. Find a parameter representation of the tangent plane at P.

f) m is the straight line which passes through $(0, 0, 0)$ and (a, a, a). Find the coordinates (x, y, z) of the point in the first octant in which m intersects the cone.

■ **1.37.** Figure 1.37 shows a rough sketch of the surface S given by $z = x^2 + y^2$, $0 \leq z \leq k^2$. Find two different parameter representations of S and do the following for each of them: (i) Draw the region D in the parameter plane which corresponds to S; (ii) find a vector $\vec{n}(P)$ which is normal to S and points to the *exterior* of S (it need not be of length 1), and determine its xyz-coordinates as functions of the parameter values corresponding to P.

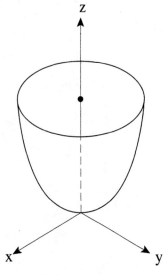

Fig. 1.37

2 | Functions of Two or More Variables

Linear functions of two and three variables

■ **2.1.** $f(x,y) = -2x - 4y + 6$.

  a) Into an xy-system draw the level curves $f(x,y) = k$ for $k = -6, 0, 6, 12$. Label the level curves.

  b) Into an xyz-system sketch the surface $z = f(x,y)$ by finding its intersection with the planes $x = 0$, $y = 0$ and $z = 0$.

■ **2.2.** $f(x,y) = Ax + By + C$. Two of the level curves of $f(x,y)$ are shown in Fig. 2.2. Find A, B, C.

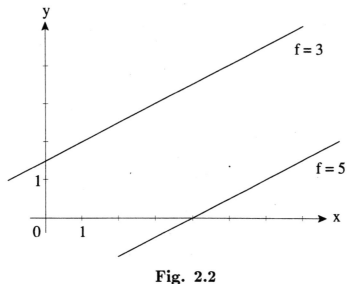

Fig. 2.2

■ **2.3.** $w = g(x,y,z) = x + 2y + z$. Sketch into an xyz-system the level surfaces $w = 8, 4, -2$ by drawing their intersections with the coordinate planes.

■ **2.4.** The plane β contains the point $C(0, 3, 5)$ and the line $x = 1+t, y = 4-3t, z = 2+t$.

  a) Find the equation of β in the form $Ax + By + \ldots$.

  b) The plane β is a surface in xyz-space, and as such it can be considered as the graph $z = p(\ldots)$ of a function $p(\ldots)$ of several variables. *How many variables?* Find the function p.

  c) β can also be considered as a level surface of a function $q(\ldots)$ of several variables, that is, β is a level surface $q = k$ where q is a function and k is a constant. Find the function q and the constant k.

Functions of two variables

■ **2.5.** Figure 2.5 shows a view of a surface $z = f(x, y)$. Your answers to a) and b) will be very rough (note that there are no units on the z-axis).

  a) Sketch into an xy-system what you think the level curves at the level of the points P and Q look like.

  b) Consider the plane β which is perpendicular to the xy-plane and which passes through the points P and Q. Sketch what you think the intersection of the surface and the plane β will look like in the plane β.

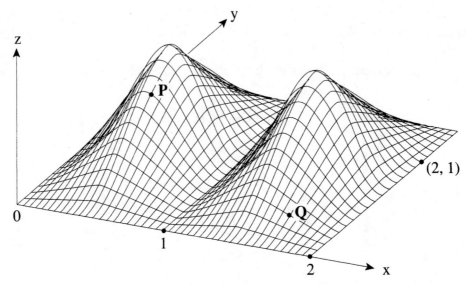

Fig. 2.5

■ **2.6.** The function $w = f(m)$ is given by its graph shown in Fig. 2.6. Note that $f(m)$ is different from 0 only for $0 < m < 4$. We use $f(m)$ to define a function $w = u(x, t)$ of two variables x, t as follows: $w = u(x, t) = f(x - t)$.

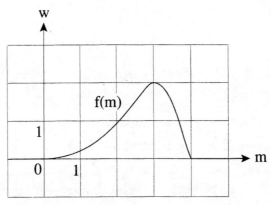

Fig. 2.6

  a) Into an xw-system sketch $w = u(x, 1)$, $w = u(x, 6)$, $w = u(x, 8)$.

  b) Into a tw-system sketch $w = u(6, t)$.

  c) Into an xt-system sketch the level curves $w = 1$ and $w = 2$.

Note. The function $w = u(x,t)$ describes a simple case of a "travelling wave" as follows: The points on the x-axis are vertically displaced as time elapses; $u(x,t)$ is the vertical displacement of the point x at time t. At time $t = 0$, the points are displaced as shown by $u(x,0) = f(x)$ (see sketch of $f(m)$). The sketches of a) show the "disturbance" at various times t, and the sketch of b) shows how the point $x = 6$ is displaced as time elapses. In the following two questions, the function $u(x,t)$ is considered as a travelling wave.

d) Find the *last* time t at which the point $x = 73$ is vertically displaced by 1 unit.

e) Find all points x which are vertically displaced by less than 0.5 unit at time $t = 212$.

■ **2.7.** The graph of the function $u = g(t)$ is shown in Fig. 2.7. We define a new function $f(x,y)$ of two variables by
$$f(x,y) = g(x^2 + y^2 - 4).$$
Describe in words the shape of the level curve $f(x,y) = 3$.

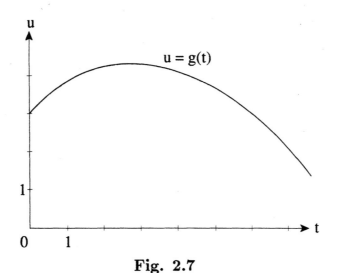

Fig. 2.7

■ **2.8.** $f(x,y) = x^3 y^3 - 9x^2 y + 16y$. Find points P and Q on the line $x = 2$ so that the following is true: *If the point (x,y) moves from P to Q, then $f(x,y)$ first decreases, then increases.*

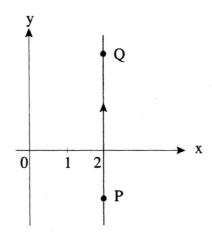

Fig. 2.8

Functions of three variables

■ **2.9.** $f(x, y, z) = (6x + 4y + 3z - 12)^{\frac{1}{3}}$.

a) Into an xyz-system draw the level surfaces

$$w = 0, \ 6^{\frac{1}{3}} \approx 1.82, \ 12^{\frac{1}{3}} \approx 2.29, \ 18^{\frac{1}{3}} \approx 2.62$$

by finding their intersections with the planes $x = 0$, $y = 0$ and $z = 0$.

b) You realized that the level surfaces are planes. Answer the following question without computation, by just looking at the pattern found in a): *The distance between the planes $w = 49235$ and $w = 49236$ is*

less than   equal to   greater than

the distance between the planes $w = 49236$ and $w = 49237$.

■ **2.10.** $w = \sin(\sqrt{x^2 + y^2 + z^2})$.

a) Find the coordinates of two points R and S on the level surface $w = 1/2$.

b) Find the equation of the level surface which contains the point $(4, 7, -3)$.

c) Describe the level surface $w = 0$, and draw part of it into an xyz-system.

■ **2.11.** $F(x, y, z) = xyz$.

a) Describe in words the level surface of $F(x, y, z)$ which passes through the origin.

b) Repeat a) for the point $(1, 0, 7)$ instead of the origin.

c) S is the level surface of $F(x, y, z)$ which passes through $(1, 1, 1)$. Write the equation of S in the form $z = g(x, y)$ and sketch the level curves $z = 1$ and $z = 433$ of S.

d) Your colleague says: "I am confused. The professor told us that functions of three variables have level *surfaces*. How come we draw level *curves* in c)?" What do you answer?

Functions of n variables

Functions $f(x, y)$ of two variables have level curves, and functions $h(x, y, z)$ of three variables have level surfaces. What about functions $F(x_1, \ldots, x_n)$ of n variables? For such functions we can define "level sets" as follows:

> Given a function $F(x_1, \ldots, x_n)$. The collection of all n-tuples $(x_1, \ldots, x_n)$ for which $F(x_1, \ldots, x_n) = constant = k$ is called the level set of the function F of level k.

We will call an n-tuple $(x_1, \ldots, x_n)$ simply a point.

■ **2.12.** Describe the level sets of the following functions of ONE variable:

$$g(t) = -t^2, \ levels \ -1, 0, 1; \quad g(t) = \sin t, \ levels \ -1, 2.$$

■ **2.13.** If $w = f(x, y, z, u)$ is a linear function of four variables:

$$w = f(x, y, z, u) = Ax + By + Cz + Du + E$$

then the level set $w = 5$ is the collection of all points (x, y, z, u) for which $f(x, y, z, u) = 5$. If D is different from zero, we can get a parameter representation of this level set by expressing u in terms of x, y, z and considering x, y, z as the parameters p, q, r:

$$x = p, \ y = q, \ z = r, \ u = \frac{1}{D}(5 - Ap - Bq - Cr - E).$$

What is the advantage of such a parameter representation? It provides easy control over all points of the level set $w = 5$ as follows. For *any* choice of p, q, r we obtain a point of $w = 5$, and any point of $w = 5$ defines uniquely parameter values p, q, r. In other words, the points (x, y, z, u) of the level set $w = 5$ are in 1:1 correspondence with the points (p, q, r) of 3-dimensional pqr-space.

a) The linear function g of six variables $x_1, \ldots, x_6$ is defined by

$$w = g(x_1, \ldots, x_6) = x_1 - 3x_2 + 2x_3 - 5x_4 + x_5 - 3x_6.$$

Find a parameter representation of the level set $w = 15$. How many parameters do you need? Call them $p_1, \ldots$.

b) $z = h(x, y) = y - 1$ (yes, there is no x). Go through the same procedure to find a parameter representation of the level set $z = 4$. Rewrite the paragraph above "What is the advantage of ..." adapted to the situation on hand.

Note. The functions whose level sets are discussed in a) and b) are both linear. For such functions it is easy to find parameter representations of level sets. For a non-linear function it is often difficult to produce parameter representations of level sets.

3 | Partial Derivatives, the Chain Rule

Partial derivatives

■ **3.1.** The surface S: $z = f(x, y)$ is symmetric with respect to the z-axis; the part $z \geq 0$ of S is shown in Fig. 3.1. C is the curve in which the plane $y = 1/2$ intersects the surface. The points A and B are in the xy-plane.

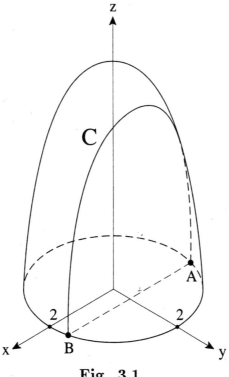

Fig. 3.1

a) A point P moves from A to B on a straight line in the xy-plane. Then the partial derivative f_x at P is (circle one):

    always positive   *first positive, then negative*   *always negative*

    first negative, then positive   *none of the preceding four*

b) Find a point Q on S, and draw it into the sketch above, at which f_x is negative and f_y is positive.

25

■ 3.2. We have the following information on a function $z = g(x,t)$ at the point $P(2,3)$:

$$g(2,3) = 5; \quad \partial g/\partial x = 1.5 \text{ at } P; \quad g(2, 2.99) = 5.04.$$

a) Estimate as well as you can: (i) $\partial g/\partial t$ at P; (ii) $g(2.008, 3)$.

b) Sketch in a tz-system what you think the graph $z = g(2,t)$ will look like for $2.7 < t < 3.2$.

c) Sketch in an xz-system what you think the graph $z = g(x,3)$ will look like for $1.7 < x < 2.2$.

■ 3.3. All that we know about a function $z = f(x,y)$ are the level curves shown in Fig. 3.3.

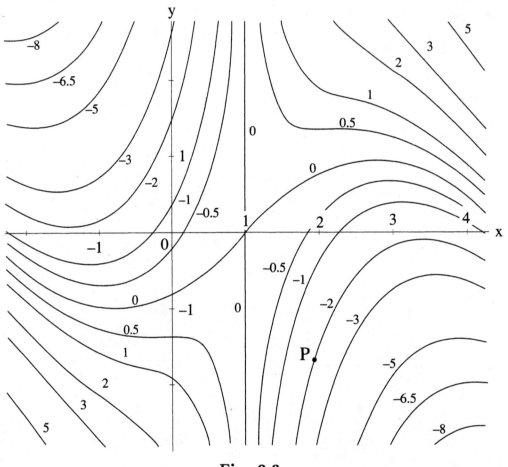

Fig. 3.3

a) Estimate as well as you can f_x and f_y at the point P.

b) Find a point Q at which both partial derivatives of $f(x,y)$ are likely to be positive.

■ 3.4. The function $g(x,y)$ is defined by

$$g(x,y) = F(\sqrt{x^2 + y^2}),$$

where $F(u)$ is a function of one variable whose graph is shown in Fig. 3.4. Find $\partial g/\partial x$ at $P(1,1)$.

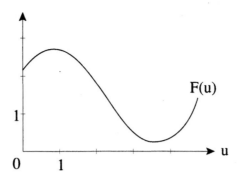

Fig. 3.4

■ **3.5.** $G(u)$ is a function of one variable. We use this function to define a function $h(x, y)$ of two variables by

$$h(x, y) = G(ax + by + c)$$

where a, b, c are constants. The partial derivatives of $h(x, y)$ can be written down in terms of the constants a, b, c and of derivatives of the function G.

Write down an expression for $h_{xx} + h_{xy} + h_{yy}$. The expression will contain a derivative of G. Show for which value this derivative has to be evaluated.

The chain rule

■ **3.6. a)** $p(x, y)$ is a function of two variables. Express

$$\frac{\partial^2 p(u + v, uv)}{\partial v \partial u}$$

in terms of u, v, and the partial derivatives p_x, p_y, ….

b) C is the level curve of level 1 of the function $f(x, y) = \sin x + \sin y$. For this C, express (i) dy/dx and (ii) d^2y/dx^2 in terms of x and y (use implicit differentiation).

■ **3.7.** The following is known about a function $f(x, y)$:

$$(f_x, f_y) = (y^2, 2xy - 16).$$

k is the straight line given by $x = 1+3t$, $y = 2+t$. Find points Q and R on k for which the following is true: *As you move a point $P(x, y)$ on k from Q to R, $f(x, y)$ first decreases, then increases.*

■ **3.8.** The following is known about the function $f(x, y, z)$:

$$(f_x, f_y, f_z) = (y, x, 1).$$

We define a new function $F(u, v)$ by

$$F(u, v) = f(2u + v, u - v, 3v^2 + 6v).$$

Find all (u, v) for which $F_u = F_v = 0$.

3.9. The function M of four variables x, y, z, w is defined by

$$M(x, y, z, w) = xy + wz,$$

and $a(t), b(t), c(t), d(t)$ are the functions

$$a(t) = 1 + 2t, \quad b(t) = 3t, \quad c(t) = 1 - t, \quad d(t) = 4.$$

Give the best estimate you can for the value of

$$\frac{M(a(1+h), b(1+h), c(1+h), d(1+h)) - M(a(1), b(1), c(1), d(1))}{h}$$

for $h = 7 \cdot 10^{-37,984}$.

4 | Linearization

Gradient and directional derivative of linear functions

■ **4.1.** The following is known about the function $f(x,y) = Ax + By + C$:

$\quad\quad grad\, f$ at $P(3,1)$ is as shown below, and $f(3,1) = 5$.

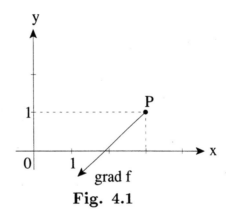

Fig. 4.1

a) Into Fig. 4.1 draw the level curve $f(x,y) = 4$.

b) Find the directional derivative of $f(x,y)$ at $P(-2,5)$ in the direction of $\vec{b} = (8,-2)$.

■ **4.2.** If $P(x,y,z)$ is a point in space, then the "steepness" of the straight line path OP can be measured by $\frac{z}{r}$ where r is the distance between the origin and the point $(x,y,0)$ (see Fig. 4.2).

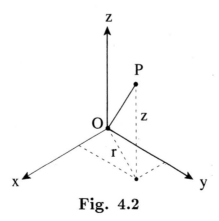

Fig. 4.2

The plane α is given by $2x - y - z = 0$; it passes through the origin. Find a point Q in α such that the path which leads from the origin to Q is the steepest possible.

■ **4.3.** In all parts of this problem, R and S will be the points $R(1,1)$, $S(3,2)$, and $\Delta \vec{r}$ is the vector which points from R to S.

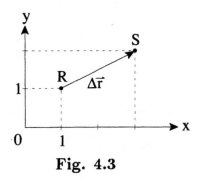

Fig. 4.3

a) We consider the function $f(x,y) = Ax + By + C$.

(i) Find the average rate of change of $f(x,y)$ between R and S, i.e., find

$$\frac{f(S) - f(R)}{|\Delta \vec{r}|}.$$

(ii) Find the directional derivative of $f(x,y)$ at R in the direction of $\Delta \vec{r}$.

(iii) You get the same answer to (i) and (ii). Would that happen again if we took different points for R and S?

b) In a) you worked with functions of *two* variables for which the average rate of change between any two points U and V is always equal to the directional derivative at U in the direction of $\overrightarrow{UV}$. Is there anything like this happening in elementary calculus? That is, are there functions of *one* variable which have a similar property? If yes, what are they? Formulate this property. Explain the property in terms of the graphs of these functions, i.e., in terms of "slope," "tangent," "secant."

■ **4.4.** This problem is based on the preceding one. We consider the function $f(x,y,z) = Ax + By + Cz + D$ and two points U, V in space. Working exactly as in **4.3.** you will find that $(f(V) - f(U))/|\overrightarrow{UV}|$, the average rate of change of $f(x,y,z)$ between U and V, is the same as the *directional derivative* of the *linear function* f at U in the direction of $\overrightarrow{UV}$. Use this fact to answer the following question.

> We do not know what A, B, C, D are in $f = Ax + By + Cz + D$, but we do know that the distance between the level planes $f = 11$ and $f = 7$ is 3 units. Find $|grad\, f|$.

Gradient and directional derivative of non-linear functions

■ **4.5.** Figure 4.5 shows some level curves of the function $z = f(x,y)$.

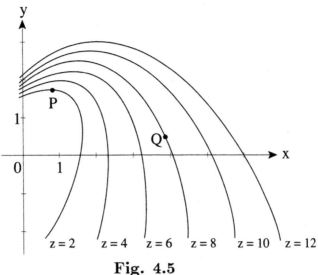

Fig. 4.5

a) Draw *grad f* at Q as well as you can.

b) By just looking at the position of P and not doing any computations, would you say that $|grad\ f|$ at Q is less than $|grad\ f|$ at P? Or greater? Or will the gradients have the same length? Explain in a complete sentence what makes you think so.

c) Find a point R at which *grad f* is parallel to $\vec{a} = (1,2)$.

■ **4.6.** We have the following information on a function $f(x,y)$: The gradient of $f(x,y)$ at $P(4,1)$ is as shown in Fig. 4.6, and $f(4,1) = 7$.

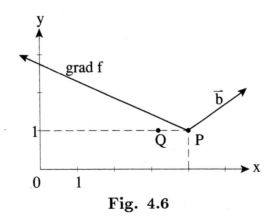

Fig. 4.6

a) Find the directional derivative of $f(x,y)$ at P in the direction of the vector $\vec{b}$ shown.

b) Find a vector $\vec{c}$ in the direction of which the directional derivative equals 4.

c) Give the best estimate you can for $f(Q)$ with Q as shown.

■ **4.7.** Figure 4.7 shows some level curves of a function $z = f(x, y)$. If P is a point on the curve C we write $\vec{t}(P)$ for the tangent vector to C at P in the direction of the arrow on C. As the point P moves from Q to R the directional derivative of $f(x, y)$ in the direction of $\vec{t}(P)$ is (circle one): *Always positive – first positive, then negative – always negative – first negative, then positive – none of these four.*

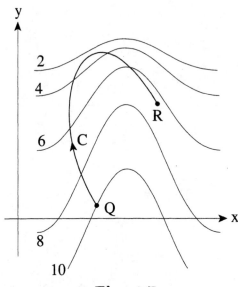

Fig. 4.7

■ **4.8.** $f(P)$ is a function defined for points P in the plane. We have the following information on $f(P)$: Its gradient at the point P_0 is of length 3 and forms an angle of $\frac{\pi}{4}$ with a fixed line g (a "reference" line), as shown in Fig. 4.8.

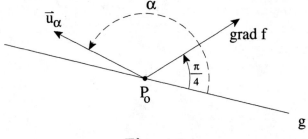

Fig. 4.8

We consider also a vector $\vec{u}_\alpha$ that forms an angle α with g, and we write $d(\alpha)$ for the directional derivative of $f(P)$ at P_0 in the direction of $\vec{u}_\alpha$. If we turn $\vec{u}_\alpha$ around P_0, i.e., if we vary α, then $d(\alpha)$ varies, too. In other words: $d(\alpha)$ is a function of α.

a) Sketch the graph $w = d(\alpha)$ for $0 \leq \alpha \leq 2\pi$.

b) By a vector $\vec{p}$ being parallel to a straight line k we mean that the line that supports $\vec{p}$ is parallel to k. Mark the following points R, S and T on the graph $w = d(\alpha)$: For R the vector $\vec{u}_\alpha$ is parallel to g; for S the tangent line to the level curve of $f(P)$ through P_0 is parallel to $\vec{u}_\alpha$; for T the vectors $\vec{u}_\alpha$ and $-grad\,f$ are parallel.

■ **4.9.** $f(x, y, z) = 2xyz + 3z^2 - y$. We look at all possible directional derivatives of $f(x, y, z)$ at the point $P(1, 1, 1)$. Find an equation (or the equations) the components of a vector $\vec{b} = (r, s, t)$ must satisfy so that the directional derivative in the direction of $\vec{b}$ is zero. Find two such vectors $\vec{b}$ which are not parallel.

■ **4.10.** We have the following information on the function $w = g(x, y, z)$ at $P(3, 7, 1)$:

$$grad\, g = (4, 4, 4) \ at \ P(3, 7, 1),\ g(3, 7, 1) = 16.$$

In addition this problem deals with two straight lines m, k which pass through P, and a point on each of them. The line m is given by $x = 3 + t$, $y = 7$, $z = 1 + 2t$, and Q is a point on m which is at a small distance b from P. The line k is perpendicular to the level surface of g at P, and R is a point on k at the same distance b from P (see Fig. 4.10).

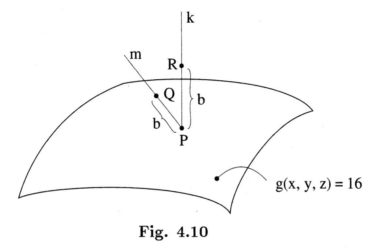

Fig. 4.10

Use directional derivatives to give estimates for $g(Q)$ and $g(R)$. The estimates will be formulas in terms of b.

■ **4.11.** The function $g(x, y, z)$ is defined by $g(x, y, z) = F(\sqrt{x^2 + y^2 + z^2})$ where $F(u)$ is a function of one variable whose graph is shown in Fig. 4.11.

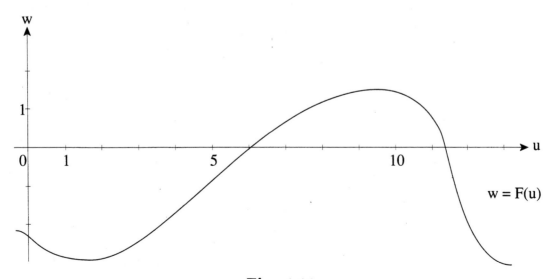

Fig. 4.11

a) Draw *grad g* at the point $S(3,9,6)$ shown in Fig. 4.11 b) (i.e., take S as initial point of the vector *grad g*).

b) A point P moves on the line RS from R to S (see below). We watch the directional derivative of $g(x,y,z)$ at P in the direction of the vector $\overrightarrow{RS}$. Is this derivative constant as P moves? Is it all the time positive? First positive and then negative? All the time negative? First negative and then positive? Or is the positive-negative pattern different from any of these? Hint: $\sqrt{x^2+y^2+z^2} = r =$ distance of (x,y,z) from $(0,0,0)$. Use r in the expression for *grad g* to read off length and direction of *grad g*.

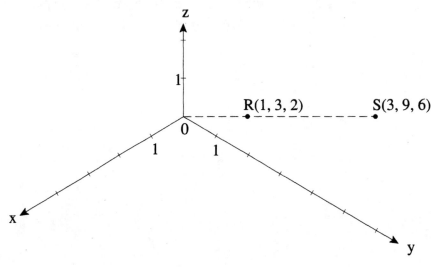

Fig. 4.11 b)

The linear approximation

Note to instructor. (1) We write $f_{lin;P_0}$ for the linear approximation of the function f at the point P_0. – (2) For functions f of one or two variables, we interpret $f_{lin;P_0}$ as the function whose graph is the tangent line or tangent plane to the graph of f at P_0 (problems 4.12, 4.13, 4.16.).

■ **4.12.** $g(t) = \sqrt{t^2+1}$. Find the function whose graph is the tangent line to the graph $u = g(t)$ at the point $(t,u) = (3, g(3))$.

■ **4.13.** $f(x,y) = (xy)^{\frac{3}{2}}$, and P_0 is the point $(x_0, y_0) = (1,4)$.

a) Find $f_{lin;P_0}$, the linear approximation of $f(x,y)$ at P_0, as the function whose graph is the tangent plane to the surface $z = f(x,y)$ at the point $(1, 4, f(1,4))$. Write the function in the form $R + Sx + Ty$.

b) Find $f_{lin;P_0}$ by using the formula $f_{lin;P_0}(P) = f(P_0) + grad\, f|_{P_0} \cdot \overrightarrow{P_0P}$. Leave your answer in the form $A + B(x-x_0) + C(y-y_0)$. Compare your answer to a).

c) Find the values (1), (2), ..., (12) below and write them into the table (the gradients are evaluated at the given (x, y)).

	$f(x, y)$	$\text{grad } f$	$f_{lin;P_0}$	$\text{grad } f_{lin;P_0}$
$(x, y) = (1, 4)$	(1):	(2):	(3):	(4):
$(x, y) = (1.02, 3.97)$	(5):	(6):	(7):	(8):
$(x, y) = (3, 5)$	(9):	(10):	(11):	(12):

d) Write down an explanation, in one or more complete sentences, for each of the following facts reflected in the table of c): (i) (1)=(3); (ii) (2)=(4); (iii) (5) is close to (7); (iv) (4)=(8)=(12) but (2), (6), (10) are different from each other; (v) (9) is very different from (11).

■ **4.14.** We have the following information on another function $G(x, y)$: The linear approximation of G at the point $(x, y) = (2, 3)$ is given by $G_{lin;(2,3)}(x, y) = 5 + x - 8y$. Find $G(2, 3)$ as well as the gradient of G at $(2, 3)$.

■ **4.15.** The temperature T, pressure p and volume V of an ideal gas are related to each other by the equation $T = kpV$ where k is a constant. We increase p_0 by Δp and V_0 by ΔV.

a) Find the corresponding change ΔT of T.

b) Use a linear approximation to give an estimate of ΔT. Is the graph of the linear approximation a curve? A surface? An object in a higher-dimensional space for which we have no visual representation?

c) Find the difference between the answer to a) and the estimate of b). Interpret this difference in terms of the surface $T = kpV$ (in pVT-space) and the graph of the linear approximation which you used in b).

■ **4.16.** We have the following information on a function $z = F(x, y)$:

$F(-1, 1) = 4$; $\text{grad } F$ at $P(-1, 1)$ is shown in Fig. 4.16.

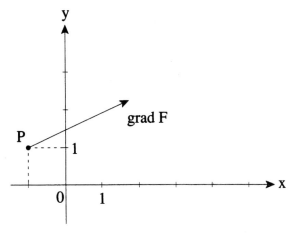

Fig. 4.16

a) Write out the formula for the linear approximation of $F(x, y)$ at $(-1, 1)$.

b) Into the system above draw the level curves $z = -2, 4, 10, 16$ of the tangent plane to the surface $z = F(x, y)$ at $(-1, 1)$.

■ **4.17.** The following is known about a function $H(x, y, z)$: The linear approximation of $H(x, y, z)$ at the point $(3, -1, 5)$ is given by the formula

$$H_{lin;(3,-1,5)}(x, y, z) = 2.5 + 7(x - 3) - 2(z - 5).$$

Let Q be the point $(7, 0, -2)$. Can you determine $H(Q)$ on the basis of the given information? What about $grad\, H$ at Q? What if Q is the point $(3, -1, 5)$?

■ **4.18.** The function $g(x, y, z, w)$ of the four variables x, y, z, w is given by

$$g(x, y, z, w) = 2xzw^3 - y^2 zw + 3.$$

a) Write out a formula for the linear approximation of g at $(1, 1, 1, 1)$.

b) Find the directional derivative of g at $(1, 1, 1, 1)$ in the direction of the vector $\vec{b} = (1, 3, 0, 4)$.

c) Find a vector $\vec{c}$ for which the following is true: If you move away from the point $(1, 1, 1, 1)$ in the direction of $\vec{c}$ then g will decrease.

d) In this part we discuss level sets of functions of 4 variables (see "Functions of n Variables" of sec. 2).

The point $(1, 1, 1, 1)$ lies on *two* level sets: (i) on the level set S of the function $g(x, y, z, w)$; (ii) on the level set S_1 of the linear approximation g_1 of g at $(1, 1, 1, 1)$.

Write out the equations of S and S_1. We can consider S_1 as an "approximation" of S. Give a parameter representation of S_1. How many parameters do you need?

Numerical computations

■ **4.19.** $f(x, y, z) = e^x z^2 \cos y + 5$. Use a linear approximation to get an estimate for

$$f(3 \cdot 10^{-427}, 10^{-427}, 1 + 2 \cdot 10^{-427}).$$

Hint: A linear approximation is only successful if it can be based at a point where we have tangible information on the function in question. Which point do you take in this problem?

■ **4.20. a)** We consider the equation

$$2xe^{5x-1} = A$$

where A is a constant. To solve this equation for x (that is, to express x in terms of the letter A) is difficult. However, one can find estimates of x for *special* values of A, viz. for values of A which are close to a value for which a solution x is known. For example, take

$A = 0.4$. Then $x = 0.2$ is a solution. Now for a value of A close to 0.4, say $A = 0.398$, one can use a linear approximation as follows. To solve

$$f(x) = 2xe^{5x-1} = 0.398$$

for x one replaces $f(x)$ with the linear approximation of $f(x)$ at $x = 0.2$, and one obtains a *linear* equation for x. Carry out this program: Write out the linear equation and solve it for x. Then use your calculator to evaluate $2xe^{5x-1}$ for the x you found and compare with 0.398.

b) We consider the system of equations

$$f(x,y) = (2\cos x + \sin x)e^y = A$$

$$g(x,y) = (-\cos x - \sin x)\sqrt{9-y} = B$$

where A and B are constants. To solve this system for x and y (that is, to express x and y in terms of the letters A and B) is difficult. However, one can find estimates for x and y for *special* values of A, B, viz. for values close to values for which solutions x and y are known. For example, take $(A, B) = (1, -3)$. Then $(x, y) = (\pi/2, 0)$ is a solution of the system

$$f(x,y) = 1, \quad g(x,y) = -3.$$

Now for values of (A, B) close to $(1, -3)$, say $(A, B) = (1.002, -2.994)$, one can use linear approximations as follows. To solve the system

$$f(x,y) = 1.002, \quad g(x,y) = -2.994$$

we replace $f(x,y)$ by linear approximations so that the system becomes a system of linear equations which can be solved easily. Carry out this program: Find the linear approximations (where do you base your approximations?) and solve the resulting system of linear equations. The (x, y) you obtain is an estimate for a solution of the system

$$f(x,y) = 1.002, \quad g(x,y) = -2.994.$$

5 | Optimization

Terminology – Note to instructor. We will use the terms maximum, extremum, etc. always with the set in mind on which the function is to be optimized. "Maximum" will mean "maximum on the domain of definition of the function," and if the function is optimized on a proper subset $\mathcal{A}$ of its domain of definition, we will say "maximum on (or in) $\mathcal{A}$." More precisely, let $f(P)$ be a real-valued function defined for points P of $\mathbb{R}^i$, $i = 1, 2, 3$, and let $\mathcal{A}$ be a subset of the domain of definition $dom\, f$ of $f(P)$. Then:

(1a) f has a **local** or **relative maximum** at $P_0 \in dom\, f$ if $f(P_0) \geq f(P)$ for all $P \in dom\, f$ such that $|\overrightarrow{PP_0}| < \epsilon$ for some $\epsilon > 0$.

(1b) f has a **maximum** at $P_0 \in dom\, f$ if $f(P_0) \geq f(P)$ for all $P \in dom\, f$. Such a maximum is also called a **global** or **absolute maximum**.

(2a) f has a **local** or **relative maximum in** or **on** $\mathcal{A}$ at $P_0 \in \mathcal{A}$ if $f(P_0) \geq f(P)$ for all $P \in \mathcal{A}$ such that $|\overrightarrow{PP_0}| < \epsilon$ for some $\epsilon > 0$.

(2b) f has a **maximum in** or **on** $\mathcal{A}$ at $P_0 \in \mathcal{A}$ if $f(P_0) \geq f(P)$ **for all** $P \in \mathcal{A}$. Such a maximum is also called a **global** or **absolute maximum in** or **on** $\mathcal{A}$.

The same applies to the various kinds of minima. — For example, consider $f(x) = x^2$ and $\mathcal{A} = [1, 2]$. Then f does not have a local minimum at $x = 1$, but it has a local minimum in $[1, 2]$ at $x = 1$.

Local and global extrema

■ **5.1.** This is a review question on local and global extrema of functions of one variable. Figure 5.1 shows the graph of a function $u = g(t)$. The points where the graph has a horizontal tangent are marked H.

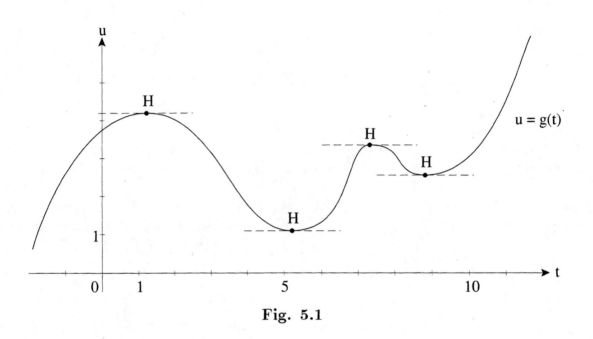

Fig. 5.1

a) Find all t for which $g(t)$ has a *local* maximum or minimum.

b) By the symbol $[a, b]$ we mean the interval $a \leq t \leq b$. (i) Find the *local* extrema of $g(t)$ in $[0, 11]$. (ii) Find the *global* extrema of $g(t)$ in $[0, 11]$.

c) Do the same as b) (i) (ii), but for the interval $[2, 4]$.

d) Find an interval $[a, b]$ such that the global maximum in $[a, b]$ is a local maximum and the global minimum in $[a, b]$ is a local minimum.

e) Find an interval $[a, b]$ such that the global maximum in $[a, b]$ is a local maximum, but the global minimum in $[a, b]$ is *not* a local minimum.

■ **5.2.** Figure 5.2 shows some of the level curves of a function $z = f(x, y)$. The level curves shown are complete in the sense that, e.g., there are no other points where $f(x,y) = 10$ other than those on the shown branches of the level curve $f = 10$. The level curves $f = -1$ are shown as small circles but are actually only points. The ellipse K in the third quadrant is not part of any level curve. Answer all questions in this problem as well as you can on the basis of the given information.

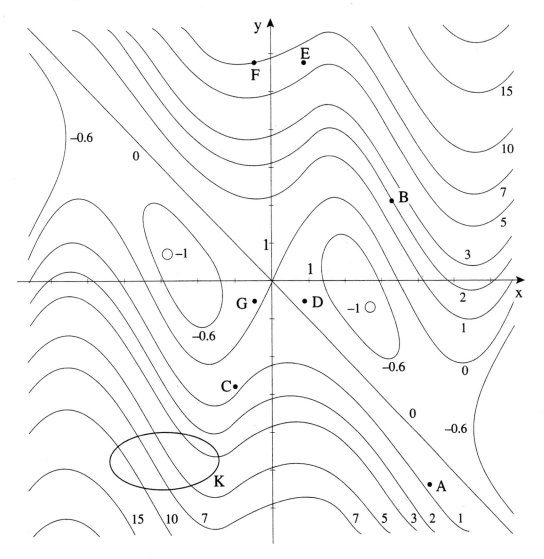

Fig. 5.2

a) Find the critical points of $f(x, y)$.

b) Classify the critical points you found in a). Are they local maxima, local minima, saddle points, or is it impossible to draw any conclusion at all?

c) Connect the points ABC by lines to obtain a triangle, and the lines $DEFG$ to obtain a rectangle. Find the global extrema of $f(x, y)$ in the solid triangle, in the solid rectangle, and in the solid ellipse K.

■ **5.3.** We have the following information on a function $f(x,y)$: Its graph $z = f(x,y)$ is the surface obtained by rotating the graph $z = g(x)$ shown in Fig. 5.3 around the z-axis.

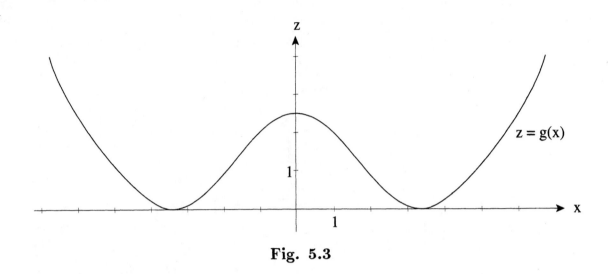

Fig. 5.3

a) (i) Find $f(2,5)$. (ii) Find all points (x_0, y_0) at which $f(x,y)$ is negative. (iii) Find all values of x_1 such that $f(x_1, 1) = \frac{3}{2}$.

b) Q_1, Q_2 are the two solid rectangles $A_1B_1C_1D_1$, $A_2B_2C_2D_2$ shown in Fig. 5.3 b). Find the maximum and the minimum of $f(x,y)$ in both rectangles. Mark the point at which $f(x,y)$ is highest, lowest in Q_1 by Max_1, Min_1 respectively, and similarly for Q_2.

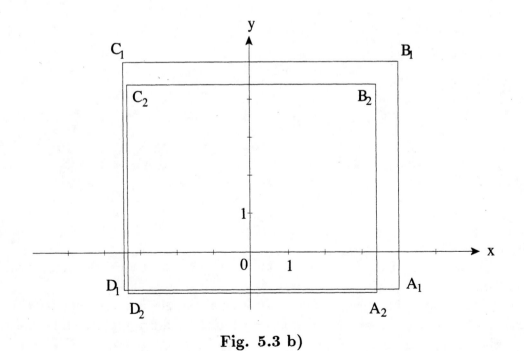

Fig. 5.3 b)

■ **5.4.** $G(t)$ is a function of one variable; its graph is shown below to the left in Fig. 5.4. We use G to define a function $f(x, y)$ of two variables by

$$f(x, y) = G(x^2 + y^2 + 5).$$

D is the shaded region shown below to the right. Find the highest and lowest values of $f(x, y)$ in D.

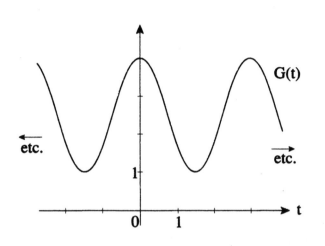
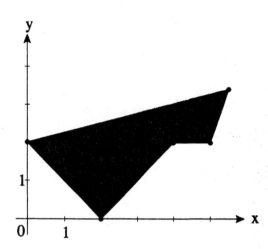

Fig. 5.4

Reasoning from the definition of local extrema

■ **5.5.** $f(x, y) = x^3 - xy^2 + 6$.

  a) Find the critical points and apply the second derivative test to them. What does the test tell you?

  b) Sketch the level curve $f(x, y) = 6$. Hint: Write $f(x, y) - 6$ as the product of three terms. The different branches of the level curve $f(x, y)$ subdivide the xy-plane into different parts. Color those parts where $f(x, y) > 6$ green, and those where $f(x, y) < 6$ red.

  c) If $(0, 0)$ is a local maximum (or minimum) then it must be possible to draw a circle of radius $\epsilon > 0$ around $(0, 0)$ so that $f(0, 0)$ is a maximum (or minimum) value of $f(x, y)$ in the interior of this circle. Can we draw such a circle? If yes, give a possible value of ϵ. If no, explain.

■ **5.6.** $f(x, y) = -3x^{16} - 2y^{28}$. Find the critical points and classify them. Hint: Do **5.5.** first.

Functions of three variables

■ **5.7.** $f(x, y, z) = 3x - y + 2z + 4$. Find the extrema of $f(x, y, z)$ on the spherical surface S given by $x^2 + y^2 + z^2 = 4$. Hint: What are the level surfaces of $f(x, y, z)$? Take the level surface $f(x, y, z) = k$ and increase k slowly from -500 to 500. How do these level surfaces "interact" with S?

44 5—Optimization

■ **5.8.** It is easily seen that $P(0, 1, 0)$ is a critical point of $F(x, y, z) = xyz$. Does $F(x, y, z)$ have a local maximum at this point? A local minimum? Neither? Explain your answer. Hint: Adapt the general idea of **5.5** to xyz-space, i.e., move away from $(0, 1, 0)$ in different directions and watch the values of $F(x, y, z)$.

Constrained extrema

Here we consider problems of the following type: Given a function $f(x, y)$ and a curve K in the xy-plane. Find the extrema of $f(x, y)$ on K. If the curve K is given in the form $G(x, y) = 0$ then the equation $G(x, y) = 0$ is called the "constraint" or "side condition," and one talks of "optimizing $f(x, y)$ under the constraint $G(x, y) = 0$."

■ **5.9.** Figure 5.9 shows some of the level curves of a function $f(x, y)$ together with five curves $A, \ldots, E$. Of the latter, all but B are thought to extend beyond what is shown of them. The curves are "oriented," i.e., each has an arrow on it to indicate a sense of direction.

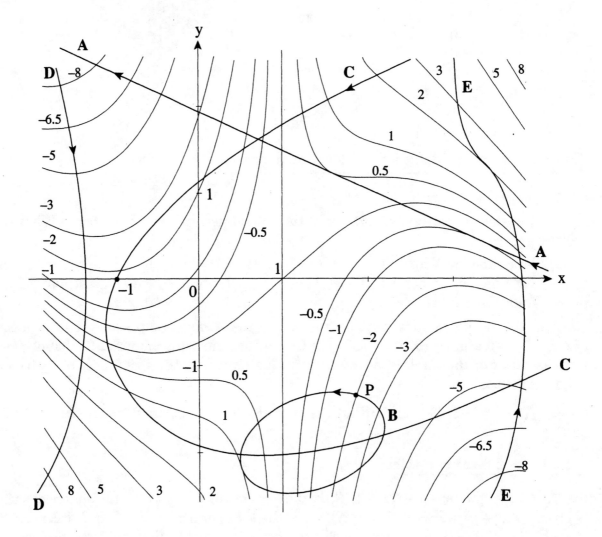

Fig. 5.9

Do the following for each curve:
(i) Move a point $P(x,y)$ over the curve in the sense of the arrow and watch the values $f(x,y)$. Are these values increasing or decreasing? Your answer will be "they first increase (or decrease), then they..., then...." On B start at P.
(ii) Move along the curve and mark all points P at which the directional derivative of $f(x,y)$ in the direction of the tangent to the curve is equal to zero. These points are the "critical points of $f(x,y)$ on the curve." Label the critical points on A by A_1, A_2, etc.
(iii) Use the information obtained in (i) to decide if the critical points of $f(x,y)$ you found in (ii) are maxima or minima or neither on the curve. Are they local or global extrema on the curve?

■ **5.10.** $f(x,y) = x^2 + y^2$. Find the critical points of $f(x,y)$ on the following curves and decide if they are extrema or not:

a) $x = 4$ **b)** $y = 1$ **c)** $y = 2x + 1$
d) $x = 1 + t$, $y = 2 + 3t$ **e)** $y = \frac{3}{x}$ **f)** $x = t^2$, $y = t^2 + t$

■ **5.11.** $f(x,y) = 2x - y + 1$. The curve K in the xy-plane is given by $x = t^2 - 2t + 5$, $y = (1/3)t^3$.

a) Find the critical points of f (note: just the critical points, and *not* the critical points on K).

b) Find the critical points of $f(x,y)$ on K and decide if they are extrema or not.

■ **5.12.** Find the critical points of $f(x,y) = x^3 + y^3$ under the constraint $x^2 + y^2 - 1 = 0$. To determine if they are extrema, proceed as follows: Draw the curve $x^2 + y^2 - 1 = 0$; mark on it the critical points you found; evaluate the function at these critical points; look at the picture and ask yourself: What *must* these points be?

■ **5.13.** $f(x,y) = x + y + 7$, and C is the curve shown in Fig. 5.13. Find the critical points of $f(x,y)$ on C and show them in the drawing. Is it possible to decide if they are extrema on C or not?

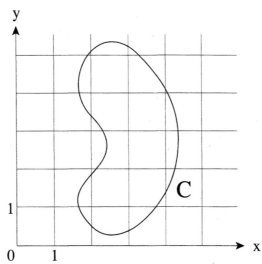

Fig. 5.13

■ **5.14.** Figure 5.14 shows some level curves of a function $z = f(x,y)$. Draw an "oriented" curve K (i.e., a curve with an arrow on it which defines a sense of direction) with points A, B, C on it with the following properties:

At A, the function $f(x,y)$ has a local maximum on K;

the point B is a critical point of $f(x,y)$ on K, and $f(x,y)$ increases before and after you cross B in the direction of the arrow;

at C, $f(x,y)$ has a local maximum on K.

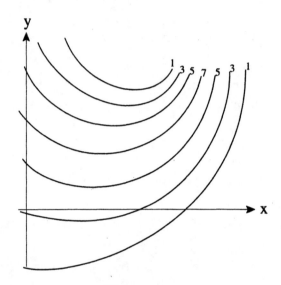

Fig. 5.14

Setting up problems with Lagrange multipliers

We consider problems of the type "Find the critical points of $f(x,y,z)$ under the constraints $h(x,y,z) = 0$ and $k(x,y,z) = 0$." If such a problem calls for the method of Lagrange multipliers, it is convenient to work as follows:

(i) Call the function $f(x,y,z)$ the "objective function" to set it apart from the equations $h(x,y,z) = 0$, $k(x,y,z) = 0$;

(ii) Write down a new function F defined by

$$F(x,y,z,\lambda,\mu) = f(x,y,z) + \lambda h(x,y,z) + \mu k(x,y,z);$$

(iii) Solve the five equations

$$F_x = 0, \ F_y = 0, \ F_z = 0, \ F_\lambda = 0, \ F_\mu = 0$$

for the five unknowns x, y, z, λ, μ (the last two equations being the constraint equations $h = 0$ and $k = 0$).

■ **5.15.** Each of the parts a),...,d) is a problem of the kind "find the critical points...." Set up each of them for solution by the Lagrange method, i.e., (i) identify and write down the objective function and the constraint(s), and (ii) write down the new function F. Do not solve the problems. Some of the objective functions have three variables, some have two, and the number of constraints varies, too.

a) The cylindrical surface $x^2 + xy + y^2 = 1$ intersects the sphere $x^2 + y^2 + z^2 = 16$ in a curve C. Find the point on C whose y-coordinate is the smallest possible.

b) R is the shaded rectangle shown in Fig. 5.15. Its upper right corner lies on the graph $y = (x - 4)^2$. Find the dimensions of R so that its area is highest.

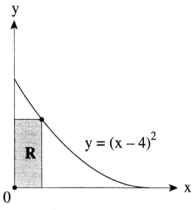

Fig. 5.15

c) Among all functions $p(x, y, z) = Ax^2 + By^2 + Cz^2$ for which $|grad\, p|_{(1,1,1)} = 1$ find those for which the directional derivative at $P(5, 0, 2)$ in the direction of $\vec{b} = (1, 3, 3)$ is highest.

d) Among all the points on the surface S given by $z = f(x, y)$ find those which are closest and farthest apart from the given point $P(a, b, c)$.

6 | Polar, Cylindrical and Spherical Coordinates

Polar coordinates (r, θ)

In problems **6.1** and **6.2** we look at the formulas which define polar coordinates

$$x = r \cos \theta, \quad y = r \sin \theta$$

as follows. Draw a rectangular system of $r\theta$-coordinates and to the right of it a system of rectangular xy-coordinates. Given a point P in the $r\theta$-plane with coordinates (r, θ), we associate with it the point Q in the xy-plane with coordinates $(x, y) = (r \cos \theta, r \sin \theta)$. For example we associate with P: $(r, \theta) = (2, \frac{\pi}{4})$ in the $r\theta$-plane the point Q: $(x, y) = (2 \cos \frac{\pi}{4}, 2 \sin \frac{\pi}{4}) = (\sqrt{2}, \sqrt{2})$ in the xy-plane. *In other words, we view the formulas of polar coordinates as a mapping from the $r\theta$-plane to the xy-plane.*

■ **6.1.** (See remark above.) Figure 6.1 shows a region K in the $r\theta$-plane and a region D_1 in the xy-plane.

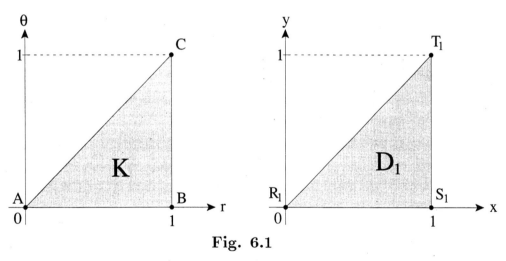

Fig. 6.1

a) Draw the region K_1 in the xy-plane which corresponds to K, that is, the region K_1 which is the image of K under the mapping from the $r\theta$-plane to the xy-plane. Show the images of the points A, B, C and label them A_1, B_1, C_1.

b) Draw the region D in the $r\theta$-plane which maps onto D_1, that is, the region D of which D_1 is the image under the mapping from the $r\theta$-plane to the xy-plane. Show the points in the $r\theta$-plane which map onto R_1, S_1, T_1 and label them R, S, T.

■ **6.2.** (See remark preceding **6.1.**) In this problem we use the following interpretation of the "area element $r\,dr\,d\theta$" of polar coordinates: *If R is a small rectangle in the $r\theta$-plane of length Δr and height $\Delta\theta$, then the area of the corresponding region in the xy-plane is approximately $r\,\Delta r\,\Delta\theta$.*

R_1 is the shaded region in the xy-plane shown in Fig. 6.2 (the sketch is not to scale). C and C^* are circles centered at the origin; C has radius $r = 6.31$ and C^* has radius $r^* = 6.44$; $\alpha = 39°15'$, and $\alpha^* = 40°30'$ (note that these angles are given in degrees).

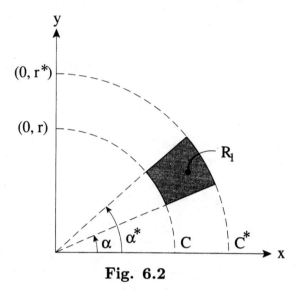

Fig. 6.2

a) Use the idea of the map from the $r\theta$-plane to the xy-plane to to find an approximation of the shaded area.

b) R_1 is a figure of elementary geometry, and there is a formula which gives the exact area A; in our case the formula gives $A = \frac{1}{2}\left((r^*)^2 - r^2\right)\Delta\alpha$. Find the area of R_1 with this formula and compare with a).

■ **6.3.** R is one of the three regions in the xy-plane shown in Fig. 6.3. Describe each of them in polar coordinates (r, θ) as follows:

Pick a θ, say $\theta = c$, and draw the ray $\theta = c$;

move out on the ray from $(0, 0)$ and call r_1 the r-value at which you enter R, and r_2 the r-value at which you leave R. In general r_1 and r_2 depend on the θ chosen, and for that reason we write $r_1 = r_1(\theta)$, $r_2 = r_2(\theta)$;

finally determine the range of θ so as to cover all of R: Move the ray $\theta = c$ by letting c increase from 0 on. Call θ_1 and θ_2 the θ-values at which you touch R the first and last time.

This produces a description of R in the form

$$r_1(\theta) \leq r \leq r_2(\theta), \quad \theta_1 \leq \theta \leq \theta_2.$$

Hint for R_3: The equation of the circle which bounds R_3 is $r = 2a\cos\theta$, $-\frac{\pi}{2} \leq \theta \leq \frac{\pi}{2}$. Note. This particular way of describing a region R in the xy-plane will be used for evaluating double integrals in polar coordinates.

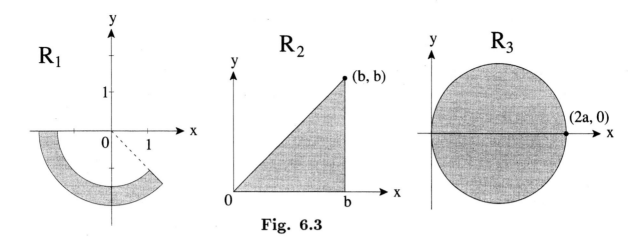

Fig. 6.3

Cylindrical coordinates (r, θ, z)

■ **6.4.** In this problem we do with cylindrical coordinates what we did with polar coordinates in **6.1**, viz. we view them as a mapping. More precisely: The formulas for cylindrical coordinates

$$x = r\cos\theta,\ y = r\sin\theta,\ z = z$$

define a mapping from $r\theta z$-space to xyz-space which assigns to the point (r, θ, z) of $r\theta z$-space the point $(r\cos\theta, r\sin\theta, z)$ of xyz-space. Figure 6.4 shows a solid K in $r\theta z$-space and a solid D_1 in xyz-space.

a) Draw K_1, the solid in xyz-space which is the image of K under the mapping defined by cylindrical coordinates.

b) Draw D, the solid in $r\theta z$-space the image of which is D_1.

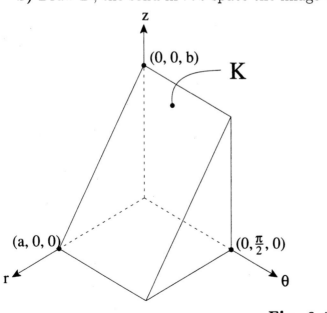
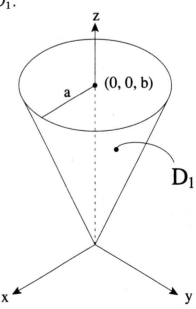

Fig. 6.4

■ **6.5.** Figure 6.5 shows three solids D. Describe each of them in cylindrical coordinates as follows:

Pick a z, say $z = c$. Intersect D with the plane $z = c$ and call the intersection W. In general the shape of W depends on the z chosen, and for that reason we write $W(z)$ for W;

make a separate drawing of $W(z)$;

describe $W(z)$ in polar coordinates as in **6.2.**: Pick a $\theta, \ldots$. In general the r-values at which you enter and leave $W(z)$ depend on the z chosen, and the same is true for the θ-values which give the range of θ for $W(z)$;

finally, determine the range of z so as to cover all of D: Move the plane $z = c$ by letting z increase. Call z_1 and z_2 the z-values at which you touch D for the first and last time.

This will produce a description of D in the form

$$r_1(\theta, z) \leq r \leq r_2(\theta, z), \quad \theta_1(z) \leq z \leq \theta_2(z), \quad z_1 \leq z \leq z_2.$$

D_1: three fourths of a heavywall pipe; D_2: straight circular cone sitting on the xy-plane; D_3: the half $z \leq 0$ of a solid sphere. *Note.* This particular way of describing a solid D in xyz-space will be used for evaluating triple integrals in cylindrical coordinates.

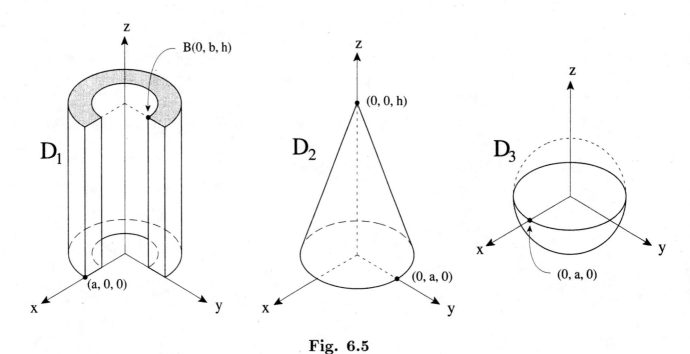

Fig. 6.5

Spherical coordinates (ρ, θ, ϕ)

■ **6.6.** In this problem we do with spherical coordinates what we did with polar coordinates in **6.1** and with cylindrical coordinates in **6.4**, viz. we view them as a mapping. More precisely: The formulas for cylindrical coordinates

$$x = \rho \cos\theta \sin\phi, \quad y = \rho \sin\theta \sin\phi, \quad z = \rho \cos\phi$$

define a mapping from $\rho\theta\phi$-space to xyz-space which assigns to (ρ, θ, ϕ) the point $(x, y, z) = (\rho \cos\theta \sin\phi, \ \rho \sin\theta \sin\phi, \ \rho \cos\phi)$ of xyz-space. Figure 6.6 shows a solid K in $\rho\theta\phi$-space and a solid D_1 in xyz-space.

a) Draw K_1, the solid in xyz-space which is the image of K under the mapping defined by spherical coordinates.

b) Draw the solid D in $\rho\theta\phi$-space of which D_1 is the image.

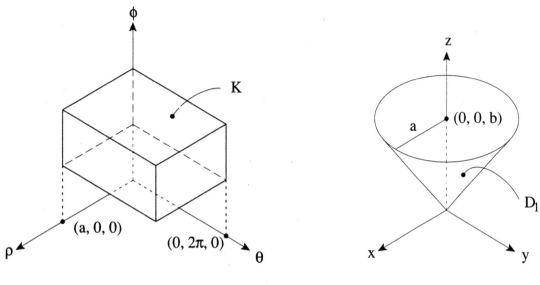

Fig. 6.6

■ **6.7.** Figure 6.7 shows three solids D. Describe each of them in spherical coordinates as follows:

Pick a θ, say $\theta = c_1$. Intersect D with the halfplane $\theta = c_1$ and call the intersection W. In general the shape of W depends on the θ chosen, and for that reason we write $W(\theta)$;

make a separate drawing of $W(\theta)$;

in $W(\theta)$ pick a value of ϕ, say $\phi = c_2$, and draw the ray $\phi = c_2$. Move out on this ray from $(x, y, z) = (0, 0, 0)$ on. Call ρ_1 and ρ_2 the ρ-values at which you enter and leave $W(\theta)$. In general ρ_1 and ρ_2 depend on the θ and ϕ chosen so far, and for that reason we write $\rho_1(\theta, \phi)$ and $\rho_2(\theta, \phi)$;

determine the range of ϕ for $W(\theta)$: Move the ray $\phi = constant$ by letting ϕ increase from 0 on. Call ϕ_1 and ϕ_2 the ϕ-values for which you touch $W(\theta)$ for the first and last time. In general these ϕ-values depend on the θ chosen, and for that reason we write $\phi_1(\theta)$ and $\phi_2(\theta)$;

finally, determine the range of θ for D: Move the halfplane $\theta = constant$ by letting θ increase from 0 on. Call θ_1 and θ_2 the θ-values for which you touch D for the first and last time.

This will produce a description of D in the form

$$\rho_1(\theta, \phi) \leq \rho \leq \rho_2(\theta, \phi), \ \phi_1(\theta) \leq \phi \leq \phi_2(\theta), \ \theta_1 \leq \theta \leq \theta_2.$$

D_1: the part $z \leq 0$ of a solid sphere; D_2: solid cap $z \geq b$ of a solid sphere; D_3: straight circular cone sitting on the xy-plane. *Note.* This particular way of describing a solid in xyz-space will be used for evaluating triple integrals in spherical coordinates.

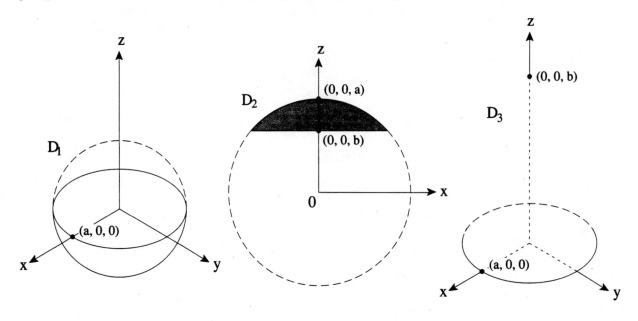

Fig. 6.7

7 Double Integrals

Note to instructor. (1) The definition of double integrals $\iint_D f\,dA$ given in many textbooks allows for fairly general regions D. If D is one of the figures of elementary geometry for which area is not an issue, such as a solid triangle or circle, the integral can be defined in terms of Riemann sums based on subdivisions of D itself. This is the approach taken in **7.2** and **7.3**. (2) Several problems belong together and should be worked one right after the other. They are: **7.1**, **7.2** and **7.3** (Riemann sums for elementary and double integrals); **7.4** and **7.5** (special Riemann sums); **7.7** and **7.8** (mean value theorem).

Riemann sums, estimates, mean value theorem

■ **7.1.** The function $u = g(t)$ is given by its graph shown in Fig. 7.1.

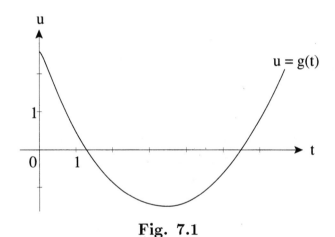

Fig. 7.1

a) Use a Riemann sum of three terms to start the process of defining

$$\int_0^6 g(t)\,dt.$$

In particular: (i) Show the way you have subdivided the interval $[0,6]$; (ii) show the t you pick in each of the three pieces; (iii) find the value of the Riemann sum.

b) Subdivide each of the three pieces of a) into two smaller pieces. Repeat the steps (i), (ii), (iii) of a) for the Riemann sum which now has six terms.

c) Find numbers m and M such that

$$m \leq \int_0^6 g(t)\,dt \leq M.$$

55

56   7—Double Integrals

■ **7.2.** T is the triangle ABC shown in Fig. 7.2, and $f(x, y)$ is the function $f(x, y) = e^{xy}$. Your answers will contain e. Leave e as a letter and do not give a numerical value.

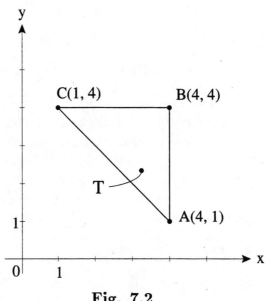

Fig. 7.2

a) Use a Riemann sum of two terms to start the process of defining

$$\iint_T f \, dA.$$

In particular: (i) Show the way you have subdivided T; (ii) show the points you have picked in each of the two pieces; (iii) find the value of the Riemann sum.

b) Subdivide each of the two pieces of a) into two smaller pieces. Repeat the steps (i), (ii), (iii) for the Riemann sum which now has four terms.

c) Find numbers m and M such that

$$m \leq \iint_T f \, dA \leq M.$$

Hint: Use the level curves of $f(x, y)$. What are they?

■ **7.3.** We continue with the triangle and the function $f(x, y)$ of **7.2**. In **7.2**, the process of defining $\iint_T f \, dA$ was *started* by subdividing T twice in a row; let us call S_1 the first subdivision and S_2 the second. To *complete* the process we have to describe a sequence of ongoing subdivisions $S_3, S_4, \ldots, S_n, \ldots$ and pass to the limit of the corresponding Riemann sums. The sequence $\ldots, S_n, \ldots$ of subdivisions cannot be completely arbitrary if we want to arrive at the integral, but must satisfy certain conditions. *What are these conditions?* Their discussion is the topic of the present exercise.

a) Let A be a region in the xy-plane (e.g., a solid triangle or rectangle). By the "diameter" of A we mean the diameter of the smallest circle which can be drawn to contain A.

(i) Find the diameter of T (of **7.2**).

(ii) Find the diameter of each of the two pieces of the first subdivision of T you used in **7.2**.

(iii) Do the same with your second subdivision of T.

If the sequence $\ldots, S_n, \ldots$ is to lead to the integral, it must have a property which involves "diameter." We call it the "diameter property," and it reads as follows: *Consider the n-th subdivision S_n. Among all the pieces of S_n there is one of maximum "diameter." If the sequence of subdivisions is to lead to the integral, then this maximum "diameter" of S_n must tend to zero as n tends to ∞.*

b) Describe a sequence of subdivisions of the triangle T which has the "diameter property." In particular: (i) Describe S_1, S_2 and S_n in words (you might or might not want to use the subdivisions you have used in **7.2**); (ii) find the maximum "diameter" of S_1, S_2 and S_n.

c) Can you think of a different sequence of subdivisions $\overline{S}_1, \overline{S}_2, \ldots$ which does *not* satisfy the "diameter" condition? Do (i) and (ii) of b) for this sequence.

■ **7.4.** $h(t)$ is the function $h(t) = 1$ for all t, and I is the interval $[0, 6]$. Subdivide I into 512 pieces, and pick a t in each of them. Form the Riemann sum. Find the value of the Riemann sum. What is the value of the corresponding integral?

■ **7.5.** $k(x, y)$ is the function $k(x, y) = 1$ for all (x, y), and T is the triangle of **7.2**. Subdivide T into 512 pieces, and pick a point (x, y) in each of them. Form the Riemann sum. Evaluate the Riemann sum. What is the value of the corresponding double integral?

■ **7.6.** The function $u = g(t)$ is given by its graph in Fig. 7.6 below left. We define a function $p(x, y)$ by

$$p(x, y) = g\left(\frac{3}{x^2 + y^2}\right).$$

D is the shaded domain shown below right. Find an estimate for the double integral of $p(x, y)$ over D, i.e., find numbers m and M such that

$$m \leq \iint_D p \, dA \leq M.$$

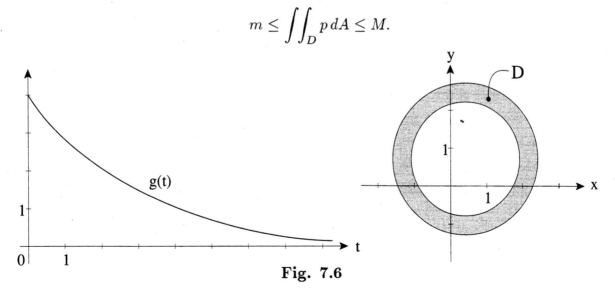

Fig. 7.6

■ **7.7.** Let $u = g(t)$ be a non-negative function, i.e., $g(t) \geq 0$. Then the mean value theorem of integral calculus says:

in **formula** language: $\int_a^b g(t)dt = g(t^*)(b-a)$ where t^* is an appropriately chosen value in the interval $[a, b]$;

in **geometry** language: The area under the graph of $g(t)$ from a to b is the same as the area of a rectangle of base $b - a$ and height $g(t^*)$ for an appropriately chosen value t^* in $[a, b]$.

Now let $g(t)$ be the function of **7.6**. Find a t^* so that

$$\int_0^5 g(t)dt = g(t^*)(5 - 0).$$

■ **7.8.** Let $z = f(x, y)$ be a non-negative function ($f(x, y) \geq 0$) and D be a domain of the xy-plane (e.g., a square or a triangle).

a) State the mean value theorem for $f(x, y)$ and D in both formula and geometry language.

b) Now we take a specific function $f(x, y) = 3 - y$, and as domain D we take the rectangle $0 \leq x \leq 1$, $0 \leq y \leq 3$. Find a point (x^*, y^*) in D such that

$$\iint_D f\, dA = f(x^*, y^*)\, A(D)$$

where $A(D)$ is the area of D. Hint: Sketch the surface $z = 3 - y$ in an xyz-system, and reason with volumes the way you reasoned with areas in **7.7**.

Setting up double integrals in cartesian and polar coordinates

■ **7.9.** In Fig. 7.9 on the following page seven regions $R_1, \ldots, R_7$ of the xy-plane are shown.

a) Describe each of them in the following way:

Pick an x, say $x = c$, and draw the line $x = c$;

walk on the line $x = c$ in the positive direction, starting with a very low value of y;

call A the point at which you enter R. The position of R depends on the $x = c$ chosen; the coordinates of A are $(x, a(x))$ where $a(x)$ is a function of x;

call B the point at which you leave R. The coordinates of B are $(x, b(x))$;

determine the range $x_1 \leq x \leq x_2$ of x so as to cover all of R.

This produces a description of R in the form $a(x) \leq y \leq b(x)$, $x_1 \leq x \leq x_2$.

b) Do a), except that you start with a line $y = c$ instead of $x = c$. You move from left to right, entering R at $S(s(y), y)$ and leaving R at $T(t(y), y)$. R is described in the form $s(y) \leq x \leq t(y)$, $y_1 \leq y \leq y_2$. Part a) and b) provide descriptions of R in the form needed to set up and evaluate double integrals over R.

c) Evaluate: (i) $\iint_{R_5} y\, dA$; (ii) $\iint_{R_6} x\, dA$.

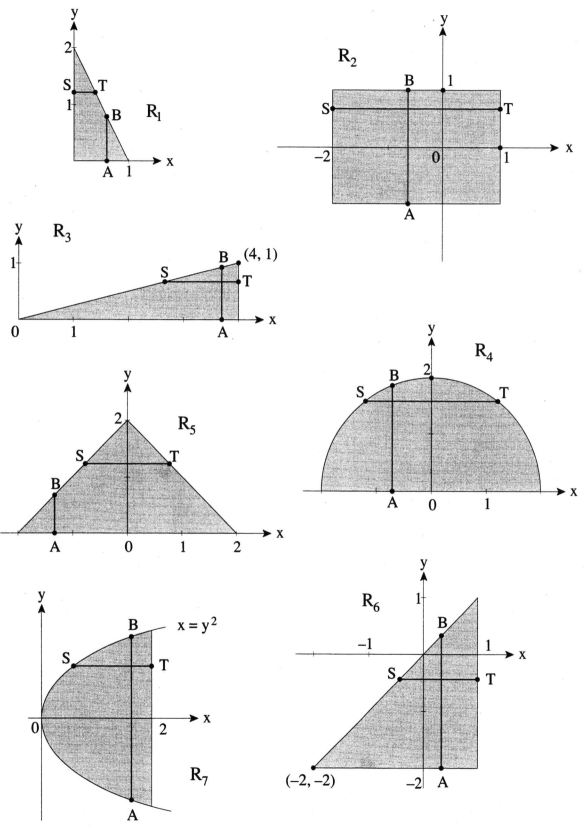

Fig. 7.9 (1) to (7)

■ **7.10.** In Fig. 7.10 D is a thin plate. Its density $\delta(P)$ at the point P is proportional to the distance of P from the line $y = 4$. We want to find the total mass of the plate.

a) Before setting up the integral for the mass of the plate decide which order of integration will make the evaluation simpler: first x, then y or the other way around.

b) Set up the integral, taking k as the factor of proportionality, and evaluate.

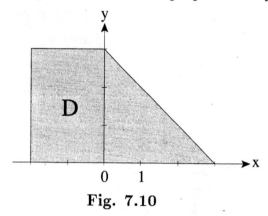

Fig. 7.10

■ **7.11.** $f(x) = \sin x + 3$, and D is the shaded domain shown in Fig. 7.11.

a) Set up an ordinary integral of one variable to find the area of D.

b) Set up a double integral to find the area of D.

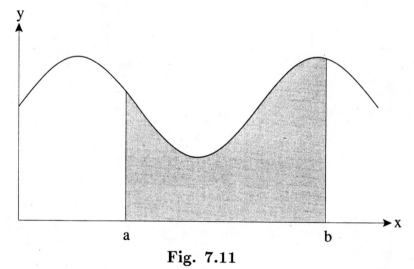

Fig. 7.11

■ **7.12.** $f(x, y)$ is a function, expressed in cartesian coordinates (x, y), and D is one of the six domains shown in Fig. 7.12. The equation of the halfcircle C (lower right) is $(x - 2)^2 + y^2 = 4$, $y \geq 0$, in cartesian coordinates and $r = 4\cos\theta$, $0 \leq \theta \leq \pi/2$, in polar coordinates.

a) We have to set up the six integrals $\iint_D f\, dA$. Decide for each D whether cartesian or polar coordinates should be used.

b) Set up the six integrals.

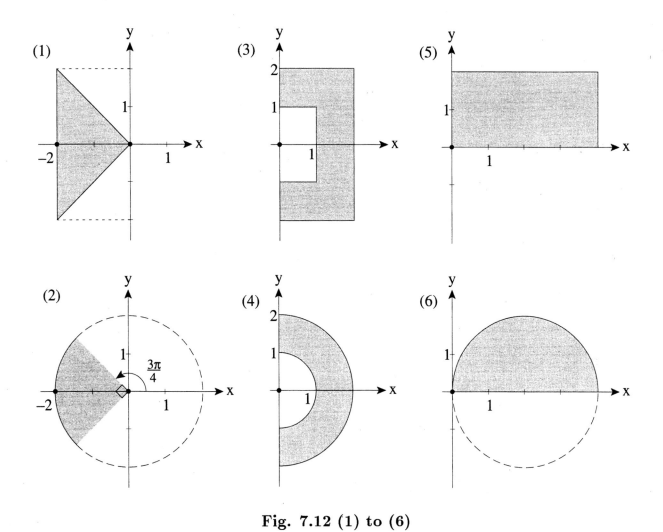

Fig. 7.12 (1) to (6)

8 Triple Integrals

Note to instructor. In **8.1** the Riemann sum is taken for elementary solids in the same way as for elementary plane figures in **7.2** (see the remarks preceding **7.1**).

Riemann sums, estimates

■ **8.1.** D is the solid circular cylinder sitting on the xy-plane, as shown in Fig. 8.1 (radius 2, height 6, axis parallel to the z-axis). The function $F(x, y, z)$ is given by

$$F(x, y, z) = \frac{1}{(x^2 + y^2 + z^2)^3}.$$

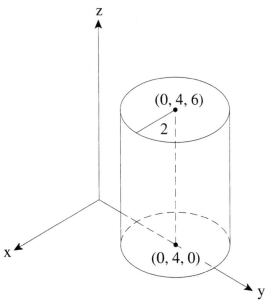

Fig. 8.1

a) Use a Riemann sum of three terms to start the process of defining

$$\iiint_D F\, dV.$$

In particular: (i) Show the way you have subdivided D; (ii) show the points you have picked in each of the three pieces; (iii) find the value of the Riemann sum.

b) Estimate the triple integral in question, i.e., find numbers m and M such that

$$m \leq \iiint_D F\, dV \leq M.$$

■ **8.2.** D is the solid ball $x^2 + y^2 + z^2 \leq 4$, and $f(x, y, z) = 3x - y + 2z + 4$. The latter is the function of **5.7**, and D is the solid sphere bounded by the spherical surface S of the same problem. Estimate (in the sense of **8.1** b))

$$\iiint_D f\, dV.$$

Setting up triple integrals in cartesian coordinates

■ **8.3.** D is the solid bounded by the surface $z = \frac{1}{4}(x - 4)^2$ and the planes $z = 0$, $y = 6$, $y = 0$, $x = 0$ (see sketch in Fig. 8.3).

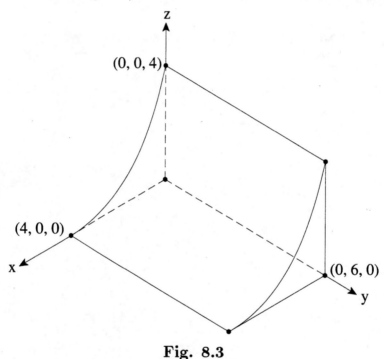

Fig. 8.3

a) Describe D as follows:

Pick an x, say $x = c$. Intersect D with the plane $x = c$ and call the intersection W. The shape of W depends on the x chosen, and for that reason we write $W(x)$ instead of W;

make a separate drawing of $W(x)$ in a yz-system;

describe $W(x)$ as a region in the yz-plane as per **7.9.**: Pick a $y, \ldots$. Note that the z-values at which you enter and leave $W(x)$ may depend on the x and y chosen so far, and the y-values for the y-range of $W(x)$ may depend on x;

finally, determine the range of x for D: Move the plane $x = c$ in the positive x-direction and call x_1 and x_2 the x-values at which you intersect D for the first and last time.

This produces a description of D in the form

$$z_1(x, y) \leq z \leq z_2(x, y), \quad y_1(x) \leq y \leq y_2(x), \quad x_1 \leq x \leq x_2.$$

b) Set up the triple integral over D of the function $f(x,y,z)$ as

$$\iiint_D f(x,y,z)\,dz\,dy\,dx,$$

i.e., by observing the following order of integration: *first with respect to z, then y, then x*.

c) Rewrite the procedure of a) for $\iiint_D f\,dx\,dy\,dz$ and set up the integral.

d) Set up $\iiint_D f\,dz\,dx\,dy$.

■ **8.4.** For each of the four solids D shown in Fig. 8.4 set up $\iiint_D f(x,y,z)\,dV$ in the given order of integration. D_1: solid tetrahedron, order of integration $dy\,dz\,dx$; D_2: solid halfcylinder of radius 2 lying on the xy-plane, order $dz\,dx\,dy$; D_3: solid prism topped by a plane, order $dx\,dy\,dz$; D_4: solid obtained by rotating the points $(0,y,z)$, $0 \le z \le \frac{1}{8}y^2$, $0 \le y \le 4$ around the y-axis, order $dz\,dx\,dy$.

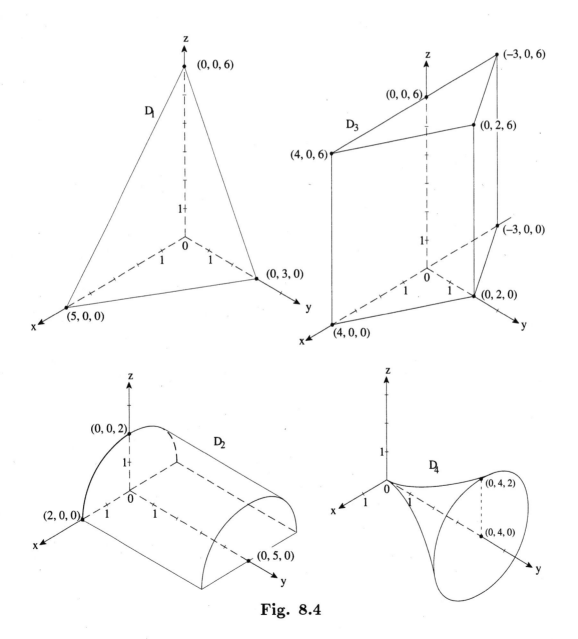

Fig. 8.4

Setting up triple integrals in cylindrical and spherical coordinates

■ **8.5.** $f(x,y,z) = xy + z$, and D is the cone D_2 of **6.5**.

a) Set up $\iiint_D f\, dV$ in cylindrical coordinates. Note that the integrand must be a function of r, θ, z.

b) Set up the same integral in spherical coordinates; here the integrand will be a function of ρ, θ, ϕ.

■ **8.6.** In this problem we view spherical coordinates as a map from $\rho\theta\phi$-space to xyz-space as in **6.6**. We also use the following interpretation of the "volume element $\rho^2 \sin\phi\, d\rho\, d\theta\, d\phi$" of spherical coordinates: *Let B be a small rectangular box in $\rho\theta\phi$-space of length $\Delta\rho$, width $\Delta\theta$, height $\Delta\phi$, and with one of its corners at the point (ρ, θ, ϕ). Then the volume of the corresponding solid B_1 in xyz-space is approximately $\rho^2 \sin\phi\, \Delta\rho\, \Delta\theta\, \Delta\phi$.*

a) D_1 is the solid in xyz-space bounded by the yz-plane, the xy-plane, the half-plane $\theta = 1.48$, the cone $\phi = 1.52$ and the spherical surfaces $\rho = 2.87, 2.93$ (see Fig. 8.6 which, for illustration's sake, is grossly out of scale). Find an approximation of the volume of D_1.

b) Set up the triple integral which gives the exact volume of D_1. Evaluate and compare with a).

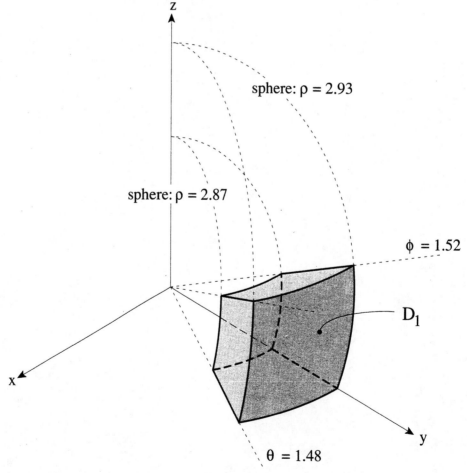

Fig. 8.6

9 | Vector Fields

Note to instructor. Let $\vec{F}(P)$ be a vector field. In order to describe $\vec{F}$ we need (1) a coordinate system to locate the points P, and (2) at each point P basis vectors $\vec{e}_1, \vec{e}_2, \ldots$ along which $\vec{F}(P)$ is decomposed. In all problems of this chapter we use cartesian coordinates for the points and $\vec{i}, \vec{j}, \ldots$ as basis vectors, except in the last section "Components in non-cartesian coordinates."

Definition, components in cartesian coordinates

■ **9.1.** $\vec{F_0}$ is a fixed vector of length 3 in space, and $\vec{F}(P)$ is the constant vector field defined by $\vec{F_0}$:

$$\vec{F}(P) = \vec{F_0} \qquad \text{for all points } P \text{ in space.}$$

a) Set up a system of xyz-coordinates and find the components $F_1(x, y, z)$, $F_2(x, y, z)$, $F_3(x, y, z)$ of $\vec{F}$.

b) Set up a system of cartesian coordinates *different from the one used in a)* and find the components of $\vec{F}$.

c) For the xyz-coordinates of a) find a scalar function $f(x, y, z)$ such that $grad\, f$ equals the vector field $\vec{F}$. How many such functions $f(x, y, z)$ can you find?

■ **9.2.** The scalar function $f(P)$ in space is defined by $f(P) = \frac{1}{2}(|\overrightarrow{AP}|^2 + |\overrightarrow{BP}|^2)$ where A and B are two fixed points in space which are 6 units apart (see Fig. 9.2).

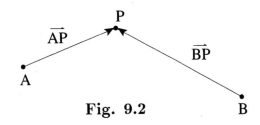

Fig. 9.2

We define a vector field $\vec{F}(P)$ by $\vec{F}(P) = grad\, f|_P$.

a) Introduce an xyz-system conveniently positioned for the points A and B. Show in a drawing of the xyz-system the points A, B. Find the components $F_1(x, y, z)$, $F_2(x, y, z)$, $F_3(x, y, z)$ of $\vec{F}$.

b) Draw into the xyz-system the field vector at the following points: (1) $K(0, 1, 0)$; (2) the midpoint M between A and B; (3) $L(2, 0, 1)$.

c) Suppose Q is a critical point of the scalar function $f(P)$. What does this mean for the field vector $\vec{F}(Q)$ at Q? Look at the work you have already done and identify a critical point of f without any further computation.

■ **9.3.** D is a plate in the xy-plane which rotates around $(0,0)$ counterclockwise with constant angular velocity; it takes 4 seconds to make one full revolution. D is shown in Fig. 9.3 at time $t = 0$ seconds with point P having coordinates $(3,0)$.

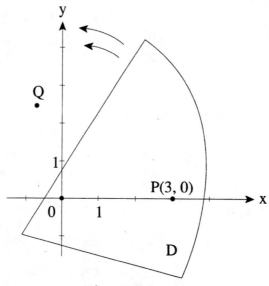

Fig. 9.3

a) As the plate rotates, the point P describes a curve C. Give a parameter representation of C with time t as parameter.

b) At time $t = 1$ the point P is at P_1. Draw P_1 and the velocity vector of P_1.

c) We define a vector field $\vec{F}$ as follows: If Q is a point in the plane, then $\vec{F}(Q)$ is the velocity vector Q would have if it were a point of a plate rotating around $(0,0)$ as does our D. Draw $\vec{F}(Q)$ for Q as shown.

d) Find the components $v_1(x,y)$ and $v_2(x,y)$ of the vector field $\vec{F}$.

■ **9.4.** In this problem, $\vec{F} = (F_1(x,y), F_2(x,y))$ will be one of the three vector fields

$$\vec{u} = \frac{1}{2}(r^2 - r)\vec{r}, \quad \vec{v} = (-y, x), \quad \vec{w} = (1/r)(y, -x) \quad (\vec{r} = (x,y), r = |\vec{r}|).$$

Answer all questions first for $\vec{u}$, then for $\vec{v}$ and then for $\vec{w}$ (nine questions to answer).

a) Into an xy-system draw $\vec{F}(P)$ for $P(1, -2)$.

b) Are there points Q where $\vec{F}(Q)$ is not defined? Where $\vec{F}(Q) = (0,0)$?

c) Let R be an arbitrary point. What is the angle between $\overrightarrow{OR}$ and $\vec{F}(R)$?

■ **9.5.** The scalar function $f(x,y,z)$ is defined by $f(x,y,z) = g(r)$, $r = |\vec{r}|$, $\vec{r} = (x,y,z)$, where g is the function of one variable whose graph is shown in Fig. 9.5. Throughout this problem we discuss the vector field $\vec{F}(P) = grad\, f|_P$.

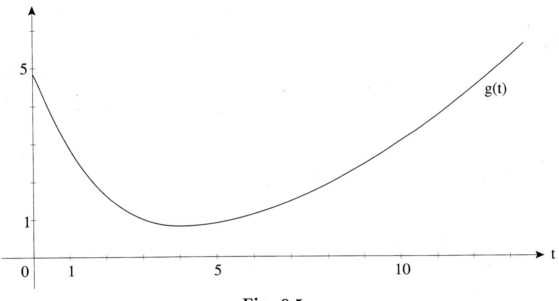

Fig. 9.5

a) Express the three components of $grad\, f$ in terms of g, r and $\vec{r}$.

b) $U(6,7,6)$ and $V(0,11,0)$ are two points on the sphere of radius 11 centered at the origin. What can you say in words about the direction of $grad\, f$ at U and V? What about $|grad\, f|$?

c) Now K and L are two points on the sphere S: $x^2 + y^2 + z^2 = m^2$. We consider a curve C on S which connects K and L (for example, the great circle through K and L). As a point P moves on C from K to L, what can you say about the direction of $grad\, f$? What about $|grad\, f|$?

d) Find two points M and N in space where the field vector is zero.

The flow lines of a vector field

■ **9.6.** $\vec{F}$ is the vector field of **9.4**. Use your answer to c) of **9.4** to describe the shape and orientation of the flow line of $\vec{F}$ which passes through the point $P(1,-2)$ (remember that $\vec{F}$ stands for the three fields $\vec{u}$, $\vec{v}$ and $\vec{w}$).

■ **9.7.** $\vec{F}$ is the vector field $\vec{F} = (0, -y)$.

a) Into an xy-system draw the field vector at the points $(-3,1)$ and $(2,-3)$.

b) What will be the general shape of the flow lines of $\vec{F}$? How will they be arranged in the plane?

c) Let C be the flow line which for $t = 0$ passes through $(4,2)$. C is described by $(x,y) = (a(t), b(t))$.

(i) Find the functions $a(t)$ and $b(t)$. Hint: The solution of the differential equation $f'(t) = -f(t)$ is of the form $f(t) = ke^{-t}$ where k is a constant.

(ii) Draw C and mark the points corresponding to $t = 0$, $t = \ln 2$, $t = 205$.

(iii) Describe in words how the point (x, y) of C moves as t increases from -10^{189} to 10^{189}.

■ **9.8.** $f(x, y) = xy$, and $\vec{F} = grad\, f$.

a) Consider a point $P(a, b)$ different from $(0, 0)$. Two curves pass through P, viz. (1) a level curve of the function $f(x, y)$, and (2) a flow line of the vector field $\vec{F}$. At which angle do these two curves intersect?

b) In an xy-system show how the flow lines of $\vec{F}$ pass through the points $(1, 1), (-1, 1), (1, -1), (-1, -1), (1, 0), (0, 1), (-1, 0), (0, -1)$ by drawing the tangents to the flowlines at these points. Hint: Sketch first the level curves of $f(x, y)$ passing through these points. Then use a).

■ **9.9.** We have the following information on a vector field $\vec{F}(x, y, z)$:

(i) $\vec{F}$ is a gradient field, i.e., $\vec{F} = grad\, f$ for some function $f(x, y, z)$.

(ii) The equation of the flow line passing through the point $(2, 4, 5)$ is $x = 2 + t^3$, $y = 4e^t$, $z = 5 + t$.

Find a vector $\vec{b}$ such that the directional derivative of $f(x, y, z)$ at $(2, 4, 5)$ in the direction of $\vec{b}$ is zero.

■ **9.10.** $\vec{F} = (F_1(x, y), F_2(x, y))$ is a vector field in the xy-plane, and C: $(x, y) = (u(t), v(t))$ is the flow line of $\vec{F}$ which for $t = 5$ passes through the point $P(a, b)$. Answer the following two questions for each of the five equations below:

(i) Are there any terms which do not make any sense at all? (Examine all ten terms.)

(ii) Does the equation express the fact that C is the flow line of $\vec{F}$ which for $t = 5$ passes through $P(a, b)$?

    A. $(F_1(a, b), F_2(a, b)) = (a(5), b(5))$.
    B. $(F_1(a, b), F_2(a, b)) = (u'(5), v'(5))$.
    C. $((\partial F_1/\partial x)|_{(a,b)}, (\partial F_2/\partial y)|_{(a,b)}) = (u'(5), v'(5))$.
    D. $(F_1(5), F_2(5)) = (u'(a, b), v'(a, b))$.
    E. $(F_1(u(5), v(5)), F_2(u(5), v(5))) = (u'(5), v'(5))$.

The tangential component of a vector field along an oriented curve

Let $\vec{F}$ be a vector field, C an oriented curve, and P a point of C. By the tangential component of $\vec{F}$ along C at P we mean the scalar projection of the field vector at P onto a tangent vector of C (in the direction of the orientation of C) at P, and we write F_{tan} for the tangential component. Thus F_{tan} is a scalar. It may be positive or negative as shown below.

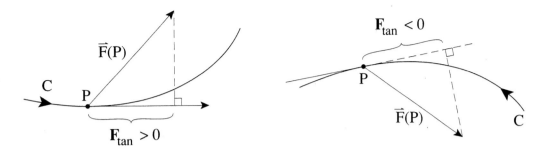

■ **9.11.** C is the oriented curve shown in Fig. 9.11 and $\vec{F}(x,y) = (-2,3)$ is a *constant* vector field. Find points P and Q on C so that F_{tan}, the tangential component of $\vec{F}$ along C, is negative at P and positive at Q. Mark the point T where F_{tan} is zero.

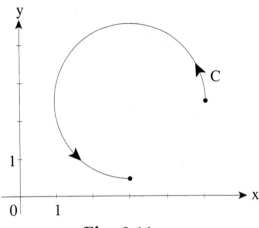

Fig. 9.11

■ **9.12.** $\vec{F}$ is the vector field $\vec{F} = (1/r)\vec{r}$ ($\vec{r} = (x,y), r = |\vec{r}|$).

a) Draw the field vector at $(2,2)$ and $(-2,1)$ into the system in Fig. 9.12.

b) C is the oriented line segment shown below. As a point P moves from A to B on C, what is the sign of F_{tan} at P, i.e., is F_{tan} always positive, always negative, or does its sign change?

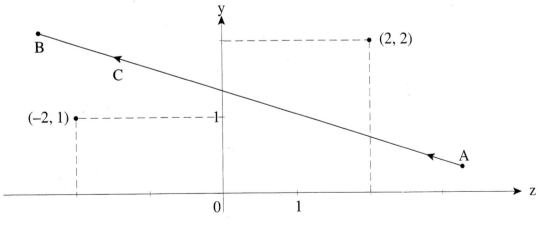

Fig. 9.12

■ 9.13. $\vec{F}$ is the vector field $\vec{F} = (x + 2y, xz, 2y)$ and C is the curve $x = 2 + t^3$, $y = 1 - t$, $z = 2t^2$. As orientation on C we take the direction of increasing t-values. Find F_{tan}. Remember that F_{tan} is a *scalar* function of t.

■ 9.14. $f(x, y) = x^2 + xy + 2y^2$, $\vec{F} = grad\ f$, and Q is the point $(1, 1)$.

   a) Into an xy-system draw $\vec{F}(Q)$.

   b) Through Q passes a flow line of the vector field $\vec{F}$ and a level curve of the function $f(x, y)$. Find the tangential component of $\vec{F}$ at Q along (i) the flow line; (ii) the level curve.

   c) Draw an oriented curve C through Q so that F_{tan} at Q along C equals 2.5 units.

Components in non-cartesian coordinates

Here we will use polar, cylindrical, or spherical coordinates for the points. As basis vectors for polar coordinates we take tangent vectors of length one to the curves $r = constant$ and $\theta = constant$ (in the obvious "increasing" directions), and similarly for cylindrical coordinates (curves $r = constant$, $\theta = constant$, $z = constant$) and spherical coordinates (curves $\rho = constant$, $\theta = constant$, $\phi = constant$).

■ 9.15. $\vec{F}$ is a constant vector field in the plane. In cartesian coordinates the components of $\vec{F}$ are $(0, -3)$.

   a) P is the point shown in Fig. 9.15. Use pencil and ruler to find the components of $\vec{F}$ at P in polar coordinates.

   b) The polar coordinates of the point P are $(r, \theta) = (4, 5\pi/6)$. Find the components of $\vec{F}$ at P in polar coordinates by *computation*.

   c) Find the formulas which give $\vec{F}$ in polar coordinates at an arbitrary point (r, θ).

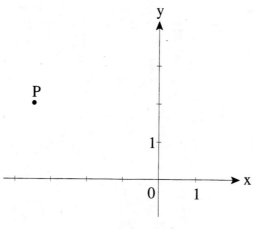

Fig. 9.15

■ **9.16.** Figure 9.16 shows some flow lines of a vector field $\vec{F}(P)$ in the plane. Find a point Q so that the components of $\vec{F}(Q)$ in polar coordinates are $(0, c)$, i.e.

$$\vec{F}(Q) = 0 \cdot \vec{e}_r + c \cdot \vec{e}_\theta.$$

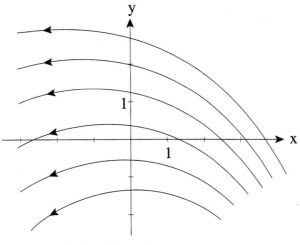

Fig. 9.16

■ **9.17.** We consider a point P which has cartesian coordinates (x, y, z), cylindrical coordinates (r, θ, z) and spherical coordinates (ρ, θ, ϕ). The point P_0 is the point on the z-axis so that PP_0 is perpendicular to the z-axis (see Fig. 9.17).

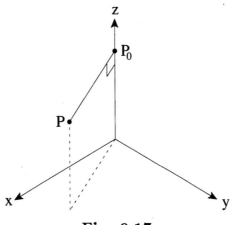

Fig. 9.17

We define a vector field $\vec{F}(P)$ by $\vec{F}(P) = 3 \cdot \overrightarrow{PP_0}$. Find the components of $\vec{F}(P)$ in (i) cartesian, (ii) cylindrical, and (iii) spherical coordinates.

10 | Line Integrals

Line integrals of scalar functions - Riemann sums, estimates

■ **10.1.** Figure 10.1 shows some level curves of a function $f(P)$ and a curve C.

a) Write out a Riemann sum of three terms to get the process of defining $\int_C f\,ds$ started. Show the points on C which you picked, and evaluate the sum.

b) The length (arc length) of C has been measured and was found to be 9.4 units. Give an estimate for $\int_C f\,ds$, i.e., find numbers k and K such that $k \leq \int_C f\,ds \leq K$.

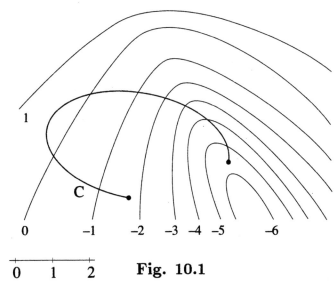

Fig. 10.1

■ **10.2.** $f(x,y) = \sqrt{x-y}$, and C is the circle of radius 1 shown in Fig. 10.2. Give an estimate for $\int_C \sqrt{x-y}\,ds$ in the sense of **10.1. b)**.

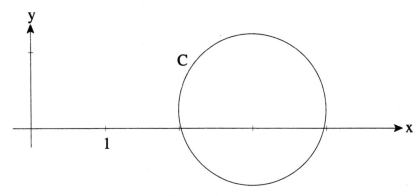

Fig. 10.2

Line integrals of scalar functions - Evaluation

■ **10.3. a)** Evaluate $\int_C f\,ds$ for $f(x,y,z) = x^2 + y^2 + z^2$ and C the line segment from $(1,3,0)$ to $(2,4,1)$.

b) Evaluate $\int_C f\,ds$ for $f(x,y,z)$ as in a) but for C the line segment from $(2,4,1)$ to $(1,3,0)$.

c) $g(t)$ is a function of one variable which we use to define a function $f(x,y,z)$ of three variables by setting
$$f(x,y,z) = g(y).$$
C is the line segment from $(4,1,7)$ to $(4,8,7)$. Set up $\int_C f\,ds$. You fall back on a notion of elementary calculus. Which one?

■ **10.4.** We consider three curves C_1, C_2 and C_3, shown in Fig. 10.4, in polar, cylindrical, and spherical coordinates, respectively:

$$C_1 : r = \theta,\ 0 \leq \theta \leq \pi$$
$$C_2 : \theta = \frac{\pi}{6},\ z = r^3 + 1,$$
$$A : (x,y,z) = (0,0,1),\ B : (x,y,z) = (\sqrt{3}, 1, 9)$$
$$C_3 : \rho = 1,\ \theta = \phi,\ 0 \leq \theta \leq \frac{\pi}{2}$$

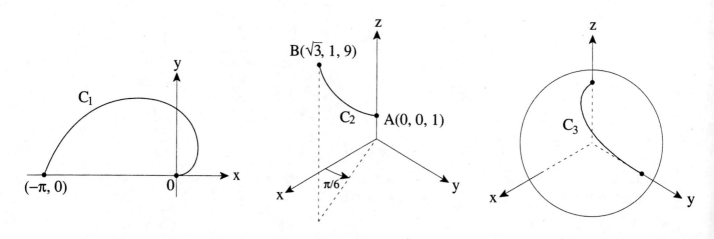

Fig. 10.4

Set up the integrals a), b), c), but do not evaluate them:

a) $\int_{C_1} xy\,ds$ **b)** $\int_{C_2} xyz\,ds$ **c)** $\int_{C_3} xyz\,ds$

Hint: If a curve C is given in polar coordinates $r = r(t)$, $\theta = \theta(t)$, then the formulas for polar coordinates provide a parameter representation in cartesian coordinates: $x = r\cos\theta = r(t)\cos\theta(t)$, $y = r\sin\theta = r(t)\sin\theta(t)$. Use the same procedure for cylindrical and spherical coordinates.

Line integrals of vector functions - Riemann sums, estimates

Note to instructor. (1) The orientation of a curve C will be given either in verbal form ("from ... to ...") or by an arrow in the sketch. If a curve is given in parametric form, the direction of increasing parameter values is taken as orientation unless otherwise mentioned. (2) If $\vec{F}$ is a vector field and C an oriented curve, we use the notation $\int_C \vec{F} \cdot d\vec{r}$ for the integral ("work integral," "circulation") of $\vec{F}$ along C. See **10.12** for the various notations in use.

■ **10.5.** $\vec{F}$ is the constant vector field $\vec{F} = \vec{F}_0$, with $\vec{F}_0$ as shown in Fig. 10.5, and C is three quarters of a circle of radius 2 units.

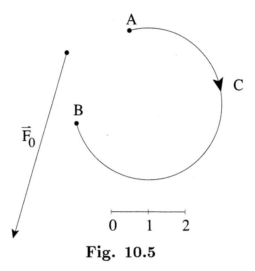

Fig. 10.5

a) Write out a Riemann sum of three terms to get the process of defining $\int_C \vec{F} \cdot d\vec{r}$ started. Show the points on C which you picked and evaluate the sum.

b) Give an estimate for $\int_C \vec{F} \cdot d\vec{r}$ (see **10.1** b) for the "estimate").

■ **10.6.** The vector field $\vec{F}$ in the xy-plane is defined by

$$\vec{F}(P) = \sin r \cdot \frac{\vec{r}}{r}, \ \vec{r} = \overrightarrow{OP}, r = |\vec{r}|,$$

and C is the line segment shown in Fig. 10.6. Give an estimate for $\int_C \vec{F} \cdot d\vec{r}$.

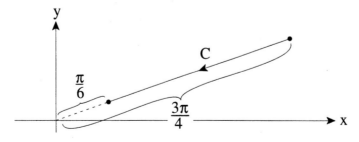

Fig. 10.6

Line integrals of vector functions - Evaluation

Note to instructor. In this section, integrals of the form $\int_C \vec{F} \cdot d\vec{r}$ are to be evaluated either straight (i.e., as integrals of a function of one variable), or by inspecting the tangential component. The topics "path independence" and "curl-test" are taken up in separate sections below.

■ **10.7.** In Fig. 10.7 C is the curve $y = \sin x$ from $x = \frac{3}{2}\pi$ to $x = 0$, and $\vec{F}$ is the vector field defined by
$$\vec{F}(x,y) = (xy, x+y).$$
Set up $\int_C \vec{F} \cdot d\vec{r}$.

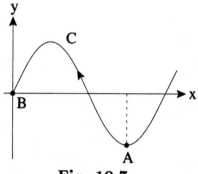

Fig. 10.7

■ **10.8.** In Fig. 10.8 C is the line segment from $A(a_1, a_2)$ to $B(b_1, b_2)$ such that $a_1 < b_1$, $a_2 < b_2$. The vector field $\vec{F}$ is defined by
$$\vec{F}(x,y) = (x, \frac{y}{2}).$$

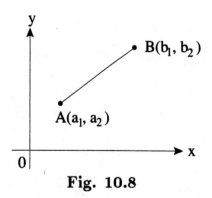

Fig. 10.8

a) Will $\int_C \vec{F} \cdot d\vec{r}$ be positive? Negative? Or is it impossible to make such a prediction unless one knows what the actual numerical values of $a_1, \ldots, b_2$ are?

b) Evaluate the integral for $A(2,4)$ and $B(8,5)$.

■ **10.9.** C is the line segment from $A(3,2,1)$ to $B(3,2,5)$. The vector field $\vec{F}$ is defined by $\vec{F}(x,y,z) = (0,0,f(z))$ where $f(z)$ is a function of one variable z. Set up $\int_C \vec{F} \cdot d\vec{r}$. You fall back on a notion of elementary calculus. Which one?

■ **10.10.** The vector field $\vec{F}$ is given by $\vec{F}(x,y) = (x^2 e^{x^2}, 2x+1)$.

a) Find the components of $\vec{F}$ at at the following points: $(3,4)$, $(3,2)$, $(3,-227)$.

b) C is the line segment from $A(1,764{,}655{,}398)$ to $B(1,764{,}655{,}378)$. Evaluate

$$\int_C \vec{F} \cdot d\vec{r}.$$

Hint: Find the tangential component along C.

■ **10.11.** The vector field $\vec{F}$ is defined by $\vec{F} = \operatorname{grad} f(r)$ where $r = \sqrt{x^2 + y^2 + z^2}$ and $f(u)$ is a function of one variable.

a) Find the field vector at the point $(3,9,-2)$ (i.e., write down the field vector in as explicit a form as possible given that the formula for the function f is not provided).

b) C is the curve whose equation in spherical coordinates is given by

$$\rho = 18, \ \theta = \pi/2, \ \phi = t \ for \ 0 \le t \le \pi/3.$$

Can you predict that $\int_C \vec{F} \cdot d\vec{r}$ is positive? Negative? Zero? Is it possible at all to make any such statement without knowing what the function $f(t)$ is?

The various notations for $\int_C \vec{F} \cdot d\vec{r}$

■ **10.12.** This problem deals with the various ways notations are used for integrals of the form $\int_C \vec{F} \cdot d\vec{r}$. The following will be considered: A vector field $\vec{F} = (F_1(x,y), F_2(x,y))$; an oriented curve C: $\vec{r}(t) = (x(t), y(t))$, $2 \le t \le 5$; the tangential component F_{tan} of $\vec{F}$ along C; the tangent vector $\vec{T} = (T_1(t), T_2(t))$ of C of length 1 in the direction of the orientation; the arc length s on C.

Which of the following expressions represent $\int_C \vec{F} \cdot d\vec{r}$ by convention or definition?

(1) $\int_C \vec{F} ds$;   (2) $\int_C F_{tan} d\vec{r}$;   (3) $\int_C F_{tan} ds$;

(4) $\int_C \vec{T} \cdot d\vec{F}$;   (5) $\int_2^5 \vec{F} \cdot \vec{r}'(t) dt$;   (6) $\int_C F_1 dx + F_2 dy$;

(7) $\int_C (F_1 + F_2) ds$;   (8) $\int_C \vec{F} \cdot \vec{T} ds$;   (9) $\int_2^5 (F_1 x' + F_2 y') dt$;

(10) $\int_C \frac{F_1 x' + F_2 y'}{\sqrt{x'^2 + y'^2}} ds$;   (11) $\int_2^5 (F_1 + F_2) ds$.

Path independence

Note to instructor. The problems in this section are designed to be worked with the following facts:

(i) $\int_C \operatorname{grad} f \cdot d\vec{r} = f(B) - f(A)$ (A initial point of C, B endpoint);

(ii) $\int_C \vec{F} \cdot d\vec{r}$ is path independent if and only if $\vec{F} = \operatorname{grad} f$.

The problems are not meant to be worked with the curl-test which is taken up in the following section.

■ **10.13.** $f(x, y, z) = x^2 + 3xyz$, and C is the spiral curve (helix) given by

$$x = 2\cos t, \quad y = 2\sin t, \quad z = 4t/\pi, \quad 0 \leq t \leq \pi/4.$$

Evaluate $\int_C \vec{F} \cdot d\vec{r}$ for $\vec{F} = \operatorname{grad} f$.

■ **10.14.** $f(x, y, z) = \cos(3x - 4y + 2z)$, $\vec{F} = \operatorname{grad} f$. In this problem you are repeatedly asked to find a curve. By this we mean, find a parameter representation and the parameter values of the initial point and endpoint (so that the curve is oriented). A curve with the required properties may or may not exist. In the latter case explain.

a) Find a curve C_1 which is not closed (i.e., its initial point is different from its endpoint) such that $\int_{C_1} \vec{F} \cdot d\vec{r} = 0$.

b) Do the same for C_2 but such that the integral is different from zero. Evaluate the integral.

c) Find a *closed* curve C_3 so that the integral is different from zero.

d) Do the same for a closed curve C_4 such that the integral is equal to zero.

■ **10.15.** In this problem, $\vec{F} = (F_1(x, y), F_2(x, y))$ will be one of the three vector fields

$$\vec{u} = \frac{1}{2}(r^2 - r)\vec{r}, \quad \vec{v} = (-y, x), \quad \vec{w} = \frac{1}{r^2}(y, -x), \quad (\vec{r} = (x, y), \ r = |\vec{r}|).$$

Answer a) for all three fields before taking up b).

a) Evaluate $\int_C \vec{F} \cdot d\vec{r}$ for the two closed curves C_1 and C_2 in Fig. 10.15.

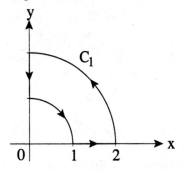 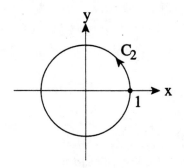

Fig. 10.15

b) Looking at the answers to a), what would be your response to the question "does $\vec{F}$ have a potential"? Circle one: yes – no – impossible to decide.

The curl-test for vector fields in the plane

Note to instructor. (1) For the moment we discuss only vector fields in the plane, and we consider the *curl* of such fields as a scalar (the *curl* of vector fields in space comes up only in Chapter 14). (2) The problems in this and the following section can be worked with the following weak version of the *curl*-test. In (I), (II) below we formulate it for $\vec{F} = (F_1(x,y), F_2(x,y))$. By "a formula in x, y makes sense for $(x, y) = (2, 3)$" is meant "if we substitute $x = 2$ and $y = 3$, we get a number."

(I) If $\text{curl}\,\vec{F}$ is different from zero, then $\vec{F}$ is not conservative.

(II) If $\text{curl}\,\vec{F}$ is zero, there are two cases:

   A. The formulas for the partial derivatives of $F_1(x,y)$ and $F_2(x,y)$ make sense for all points of the xy-plane. Then $\vec{F}$ is conservative.

   B. There is at least one point where the formula for one of the partial derivatives of $F_1(x,y)$ or $F_2(x,y)$ does not make sense. Then the test is inconclusive, that is, $\vec{F}$ may or may not be conservative.

This version of the *curl*-test also works for vector fields in space.

■ **10.16.** $\vec{F}$ is the vector field of **9.3** (rotating plate). Is $\vec{F}$ conservative? If yes, find a potential. If no, explain.

■ **10.17.** $\vec{F}$ is the vector field $\vec{F} = r^2 \cdot \vec{r}$ ($\vec{r} = (x, y)$, $r = |\vec{r}|$). Is this vector field conservative? If yes, find a potential. If no, explain.

■ **10.18.** $\vec{F}$ is the vector field $\vec{w}$ of **10.15**. Apply the *curl*-test to $\vec{F}$. Is $\vec{F}$ conservative or not? Go back to **10.15** to answer that question.

■ **10.19.** $\vec{F}$ is the vector field $\vec{F} = (\frac{x}{r}, \frac{y}{r})$ (r as in **10.17**).

   a) Apply the *curl*-test to $\vec{F}$.

   b) C is the path from $(1, 0)$ to (x, y) shown in Fig. 10.19. Evaluate $\int_C \vec{F} \cdot d\vec{r}$. The result will be an expression in terms of the coordinates (x, y) of the endpoint of C. Take the partial derivatives of this expression with respect to x and y.

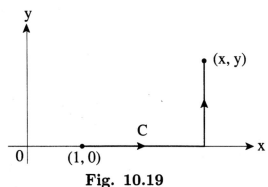

Fig. 10.19

   c) Is $\vec{F}$ conservative? If yes, find a potential. If no, explain.

- **10.20.** The *curl*-test as given at the beginning of this section deals with four situations:
  1. $\operatorname{curl} \vec{F} \neq 0$;
  2. $\operatorname{curl} \vec{F} = 0$, and the formulas for the partial derivatives of $F_1(x,y)$ and $F_2(x,y)$ make sense for all points;
  (3a) $\operatorname{curl} \vec{F} = 0$, there is at least one point where the formula for one of the partial derivatives of $F_1(x,y)$ or $F_2(x,y)$ does not make sense, $\vec{F}$ conservative;
  (3b) $\operatorname{curl} \vec{F} = 0$, there is at least one point where the formula for one of the partial derivatives of $F_1(x,y)$ or $F_2(x,y)$ does not make sense, $\vec{F}$ not conservative.

  Which of these situations is illustrated by each of the four problems **10.16**...**10.19**?

- **10.21.** The flowlines of the vector field $\vec{F}$ have the shape of parallel curves $y = f(x) + c$ where c is a constant (see Fig. 10.21). The equation of the flowline passing through the point (a, b) is
$$x = t, \quad y = f(t) + b - f(a)$$

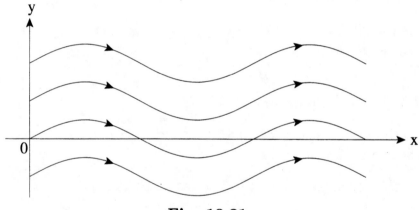

Fig. 10.21

  a) Find the components $(F_1(x,y), F_2(x,y))$ of $\vec{F}$.

  b) Give examples of functions $f(t)$ so that (i) $\vec{F}$ is conservative; (ii) $\vec{F}$ is not conservative.

Confusion surrounding the curl-test

- **10.22.** $\vec{F} = (xy, 0)$, and C is the circle $x^2 + y^2 = 1$ in the xy-plane.

  a) Integrate $\vec{F}$ once around C in a counterclockwise direction.

  b) Find $\operatorname{curl} \vec{F}$.

  c) Your inquisitive colleague says: "Wait a minute. There is something wrong. The integral around C is zero. Therefore it follows from the *curl*-test that $\operatorname{curl} \vec{F}$ must be zero. But the *curl* is *not* zero." What do you answer?

■ **10.23.** $\vec{F} = (1/r^2)(-y, x)$ (see **10.15**).

a) Find $\text{curl } \vec{F}$.

b) Your colleague is really a pest and says: "In **10.15** we computed the integral of this vector field around the unit circle, and we found the integral to be different from zero. But we just saw in a) that $\text{curl } \vec{F}$ is zero. Doesn't the *curl*-test tell us that the integral must be zero because the *curl* is zero ? What's going on here?" What do you say?

11 | Flux and Circulation in the Plane, Green's Theorem

The normal component of a vector field along a curve in the plane

We consider the curve K shown in Fig. 11(i) below left. At point Q on K we can draw vectors normal to K in two directions, viz. in the directions of $\vec{a}$ and $\vec{b}$.

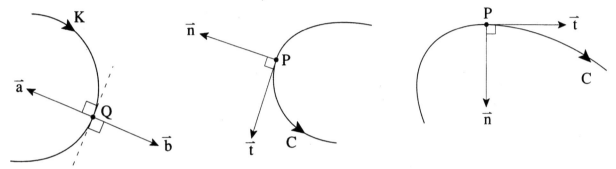

Fig. 11 (i), (ii) and (iii)

We need procedures to choose one of these two normal directions. This is done for *oriented* curves (curves with a sense of direction on them) as follows. Consider an oriented curve C, and let $\vec{t}$ be a tangent vector of C at point P in the direction of the orientation. As normal vector $\vec{n}$ of C we pick one to the *right* of $\vec{t}$. In other words, if you turn $\vec{n}$ by $\frac{\pi}{2}$ in a counterclockwise sense you obtain a vector parallel to $\vec{t}$, as shown in the two examples Fig. 11(ii) and 11(iii) above.

In addition we consider a vector field $\vec{F}$. The normal component of $\vec{F}$ along C at P is the scalar projection of $\vec{F}$ onto a normal vector chosen in the way just described. As with the tangential component of a vector field along a curve, the normal component is a scalar and can be positive or negative. Note that the way we just chose a normal to a curve is different from the definition of the principal normal in connection with the curvature of a curve.

■ **11.1.** Take the oriented curve C and the vector field $\vec{F}$ of problem **9.11** (C is three quarters of a circle). The normal component F_n of $\vec{F}$ along C is a function $v_n(s)$ of s where s is the arclength on C counted from the initial point on. Make a "rough" sketch of $v_n(s)$ (a "rough" sketch of the graph of a function $h(x)$, $x_0 \le x \le x_1$, shows the sequence in which $h(x)$ is positive/negative and increasing/decreasing as x increases from x_0 to x_1).

■ **11.2.** $\vec{F} = (x + 2y, 2x)$, C: $x = t + t^3$, $y = 1 - t$, oriented by increasing t-values. Use formulas to find the normal component of $\vec{F}$ along C at the point where $t = 1$.

Flux across a curve in the plane

The flux of $\vec{F} = (F_1, F_2)$ across the curve C for which a normal has been chosen is defined as $\int_C F_n\, ds$ where F_n is the normal component of $\vec{F}$ along C. If C: $\vec{r}(t) = (x(t), y(t))$ is oriented in the direction of increasing t-values, the flux across C is evaluated as $\int_C -F_2\, dx + F_1\, dy$.

■ **11.3.** $\vec{F}$ is the constant vector field $\vec{F} = \vec{F}_0$ shown in Fig. 11.3.

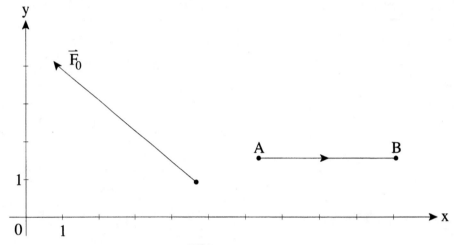

Fig. 11.3

a) Use pencil and ruler to find the flux of $\vec{F}$ across the directed line segment AB (recall our convention about normals).

b) Draw a directed line segment KL of length 2 cm for which the flux is 6 units.

■ **11.4.** C is the straight line segment shown in Fig. 11.4. Find the flux across C of the following two vector fields:

(i) $\vec{F} = grad\,(7x + 5y)$;   (ii) $\vec{F} = (4x, -2y)$.

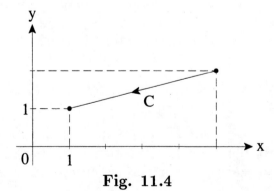

Fig. 11.4

■ **11.5.** $\vec{F}$ is the vector field defined by $\vec{F} = (0, f(x))$ where $f(x)$ is a function the graph of which is shown in Fig. 11.5. R is the solid triangle OPQ, and C is its boundary.

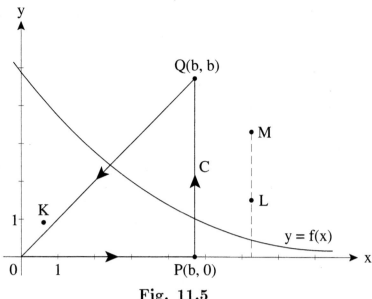

Fig. 11.5

a) *Draw the field vector at the points K, L, M.*

b) We look at the flux of $\vec{F}$ across each of the three sides of the triangle. For each of these sides predict if the flux is positive, or zero, or negative, or if it is impossible to make any prediction unless we know the formula for $f(x)$. What do you think: Will the flux across the whole of C be positive, or zero or negative? Or is it too close to call?

c) Push the evaluation of the flux across C as far as you can on the basis of the given information.

■ **11.6.** C is the circle $x^2 + y^2 = b^2$, oriented clockwise. $\vec{F}$ and $\vec{G}$ are the vector fields

$$\vec{F} = (\frac{x}{r}, \frac{y}{r}), \quad \vec{G} = (-\frac{y}{r}, \frac{x}{r}), \quad r = \sqrt{x^2 + y^2}.$$

Find the flux across C for both vector fields. Hint: Make a sketch of C and reason with the normal components along C.

Green's theorem for flux and circulation in the plane

Note to instructor. We interpret Green's theorem in terms of circulation and flux of a vector field in the plane so that there are two equivalent formulations:

$$\int_C F_{tan}\,ds = \iint_D \text{curl}\,\vec{F}\,dA, \quad \int_C F_n\,ds = \iint_D \text{div}\,\vec{F}\,dA.$$

In these equations F_{tan} is the tangential component of $\vec{F}$ along the oriented curve C, $\text{curl}\,\vec{F}$ is a scalar (see the note preceding **10.16**), F_n is the normal component along C, and D is the domain whose boundary is C. For Green's theorem to hold, $\vec{F}$, C and D must satisfy certain conditions. The vector fields, curves and domains discussed below either satisfy these conditions or are tacitly assumed to do so.

■ **11.7.** Go back to **11.5**: Use Green's theorem to find the circulation of $\vec{F}$ along C and the flux across C. Hint for the circulation: First integrate with respect to x.

■ **11.8.** $\vec{F}$ is the vector field in the plane defined by $\vec{F} = (ax + by + c, mx + ny + p)$ where a, b, c, m, n, p are constants. Can you find values for these constants, not all equal to zero, so that for any closed curve C both the flux across C and the circulation along C are zero? If yes, find such values. If no, explain.

■ **11.9.** $f(x, y) = e^x \sin y$.

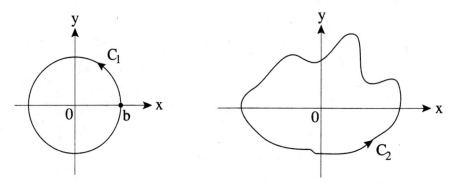

Fig. 11.9

a) Find the flux of $\vec{F} = grad\, f$ across the circle C_1 and the closed curve C_2 (Fig. 11.9).

Comment on notation. In the flux integrals of a) the integrand $(grad\, f)_n$, the normal component of $grad\, f$ along the curves, is the directional derivative of f in the direction of the normal to the curves. This derivative is often abbreviated as $\frac{\partial f}{\partial n}$.

b) Find the circulation of $\vec{F} = grad\, f$ along C_2 (think it through and do not use your pencil for at least 30 seconds).

■ **11.10.** $\vec{F}$ is the vector field in the plane defined by $\vec{F} = (k/r^2)\vec{r}$, $\vec{r} = (x, y)$, $r = |\vec{r}|$, k constant > 0. In Fig. 11.10(i) below the arcs AD and BC are parts of circles.

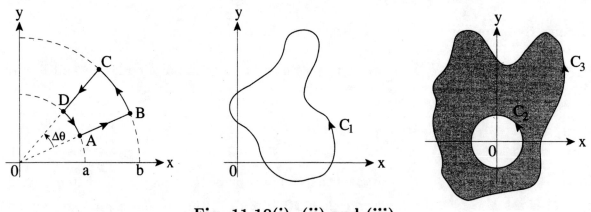

Fig. 11.10(i), (ii) and (iii)

a) Find the flux of $\vec{F}$ across the arc DA in Fig. 11.10(i) with the normal along DA pointing towards the origin.

b) Use the result of a) to find the flux across the closed curve $ABCD$.

c) Is it OK to apply Green's theorem to find the flux across $ABCD$? Whatever your answer is, explain; if yes, do it.

d) Let C_1 be a closed curve which does not contain $(0,0)$ in its interior (Fig. 11.10(ii)). Use your reasoning in c) to find the flux across C_1.

e) C_2 is the circle $x = b\cos t$, $y = b\sin t$, $0 \leq t \leq 2\pi$ (Fig. 11.10(iii)). Is it OK to apply Green's theorem to find the flux of $\vec{F}$ across C_2? Whatever your answer is, explain. Find the flux.

f) C_3 is a closed curve which contains the circle C_2 in its interior. Apply Green's theorem to the shaded region between C_3 and the circle C_2 to find the flux of $\vec{F}$ across C_3.

g) Summarize your work: "The vector field $\vec{F} = (k/r^2)\vec{r}$ is defined for all points except .... Its divergence is .... The flux across any simple closed curve which does not contain $(0,0)$ is .... The flux across any simple closed curve which does contain $(0,0)$ is ...."

■ **11.11.** Throughout this problem, $\vec{F}$ is the vector field

$$\vec{F} = \left(\frac{x}{2}, \frac{y}{2}\right).$$

a) Find the flux of $\vec{F}$ across the closed curve C shown in Fig. 11.11.

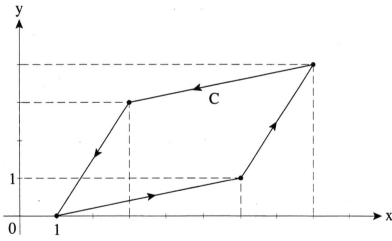

Fig. 11.11

b) It is assumed that you worked a) by using Green's theorem (if you have not done so, do it now). Complete the following sentence: "The simple closed curve C is the boundary of a domain D in the xy-plane. Then the surface area of D can be computed by the line integral ...."

c) D is the domain of **7.11** bounded by $x = 0$, $y = a$, $y = b$, $y = \sin x + 3$. In **7.11** you set up an *ordinary* integral and a *double* integral to compute the area of D. Use b) of the present problem **11.11** to set up a *curvilinear* integral to find the area of D.

Invariant definition of curl and divergence in the plane

■ **11.12.** $\vec{F}$ is the vector field in the plane defined by $\vec{F} = (y^2, y)$. D is the solid square in Fig. 11.12, and C is its boundary.

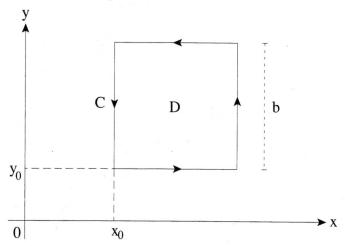

Fig. 11.12

In the course of this problem you have to find the circulation along C and the flux across C. There are two ways of doing that: Either you use standard procedures ("find a parameter representation ..."), or you reason directly with the normal and tangential component of $\vec{F}$. The latter method works extremely well in this problem because each part of C is parallel to the x-axis or the y-axis. Therefore the normal or tangential component of $\vec{F}$ is either equal to the x-component of $\vec{F}$ (which is y^2) or to the y-component (which is y).

a) Find the circulation along C. If you divide this circulation by the area of D you get "circulation per area." Now let b tend to zero so that the square shrinks to the point (x_0, y_0). What is the limit of this "circulation per area"? Confirm your answer by finding $\operatorname{curl} \vec{F}$ at (x_0, y_0).

b) Find the flux of $\vec{F}$ across C. If you divide this flux by the area of D you get "flux per area." Now let b tend to zero as in part a). What is the limit of this "flux per area"? Confirm your answer by finding $\operatorname{div} \vec{F}$.

■ **11.13.** $\vec{F}$ is the vector field in the plane defined by $\vec{F} = (2x^2 y, x^2 + y^2)$, and M is one of the three points $(-0.5, 1)$, $(2, 2)$, $(1, 1)$. C is a circle of radius 10^{-3} units, centered at M, and oriented counterclockwise.

Give an estimate for the circulation of $\vec{F}$ along C and the flux across C (six questions to answer).

12. Surface Integrals, Flux Across a Surface

Note to instructor. Problems **12.1** and **12.9** deal with Riemann sums for surface integrals over a circular cylinder S. The Riemann sums are meant to be based on subdivisions of S itself.

Surface integrals of scalar functions - Riemann sums, estimates

■ **12.1.** S is the cylindrical surface $x^2+(y-5)^2=4$, $0 \leq z \leq 6$ (see Fig. 12.1), and $f(x,y,z)$ is the function
$$f(x,y,z) = (x^2+y^2+z^2)^{-1}.$$

a) Set up a Riemann sum of three terms to get the process of defining
$$\iint_S f\,dA$$
started. In particular: (i) Show the way you have subdivided S; (ii) show the points which you have picked in each of the three pieces; (iii) find the value of the Riemann sum.

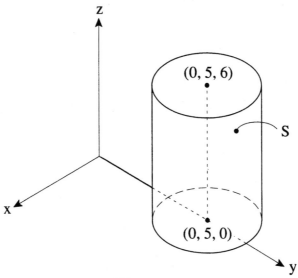

Fig. 12.1

b) Give an estimate of $\iint_S f\,dA$, i.e., find numbers k and K so that
$$k \leq \iint_S f\,dA \leq K.$$

■ **12.2.** We consider the function $f(x,y,z) = 3x-y+2z+4$ and the surface $S: x^2+y^2+z^2 = 4$ of **5.7**. Use the results of **5.7** to give an estimate of $\iint_S f\,dA$ in the sense of the preceding problem.

Surface integrals of scalar functions - Evaluation

■ **12.3.** S is the spherical shell $x^2 + y^2 + z^2 = 9$, and the function $f(x, y, z)$ is defined by

$$f(x, y, z) = H(\sqrt{x^2 + y^2 + z^2})$$

where $H(t)$ is a function of one variable given by its graph shown in Fig. 12.3. Find $\iint_S f\, dA$.

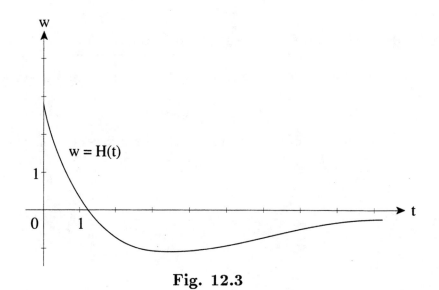

Fig. 12.3

■ **12.4.** S is the parallelogram in space with vertices $A(1, 0, 1)$, $B(4, 3, 5)$, $C(6, 4, 6)$ and $D(3, 1, 2)$ (Fig. 12.4 is not to scale). The function $f(x, y, z)$ is defined by

$$f(x, y, z) = 4y + z.$$

Find $\iint_S f\, dA$.

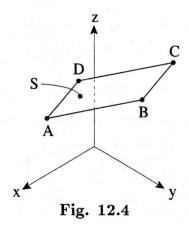

Fig. 12.4

■ **12.5.** S is the fourth of the cylindrical surface defined by

$$y^2 + z^2 = 4, \ 0 \leq x \leq 3, y \geq 0, z \geq 0,$$

(see Fig. 12.5). The scalar field $f(P)$ is defined by

$$f(P) = |\overrightarrow{OP}|^2.$$

Find $\iint_S f \, dA$. Hint: Use x and the θ of polar coordinates in the yz-plane as parameters, i.e., $x = x, y = 2\cos\theta, z = 2\sin\theta$.

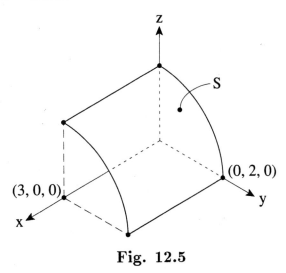

Fig. 12.5

The normal component of a vector field on a surface

We consider the surface S shown below. At point P on S we can draw vectors normal to S in two directions, viz. in the directions of the vectors $\vec{a}$ and $\vec{b}$.

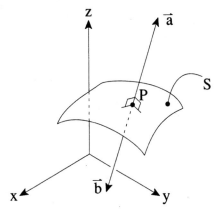

From now on, whenever we consider a surface we will choose for it one of the two normal directions, and we will say which one. For example, on the sphere $x^2 + y^2 + z^2 = 1$ we may choose the normal pointing towards the origin, or for the plane $\vec{r} = \vec{a} + u\vec{b} + v\vec{c}$ we choose the vector $\vec{b} \times \vec{c}$ as normal.

Now we consider a surface S with a normal chosen and a vector field $\vec{F}$. The normal component of $\vec{F}$ at point P on S is the scalar projection of the field vector at P onto the normal chosen for S. The normal component is a scalar and can be positive or negative.

■ **12.6.** The points $A(1,0,1)$, $B(4,3,5)$ and $C(3,1,2)$ define a plane α. As normal we take $\vec{AB} \times \vec{AC}$. In this problem we consider different vector fields, but always the same plane α.

a) $\vec{F}$ is the constant vector field $\vec{F} = (3,1,5)$. Find the normal component F_n of $\vec{F}$ on α.

b) $\vec{F}$ is the vector field $\vec{F} = r^{-2}\vec{r}$ where $\vec{r} = (x,y,z)$, $r = |\vec{r}|$. Find the normal component F_n of $\vec{F}$ at points A and C of α.

c) We take $\vec{OP} = \vec{OA} + u\vec{AB} + v\vec{AC}$ as a parameter representation of α. Find the normal component F_n of $\vec{F} = (5y, 7, x+z)$ at the following points of α: (i) Q: $(u,v) = (1,4)$; (ii) R: (u,v) (the "generic point" of α). You realize in (ii) that the normal component is a scalar function of the parameters of the surface in question.

■ **12.7.** S is the surface $z = 1 + x^2 + y^2$, and $\vec{F} = grad(r^{-2})$ with r as in **12.6** b). The normal points away from the z-axis. Find the normal component F_n at the point $(\frac{1}{2}, \frac{1}{2}, \frac{3}{2})$ on S (the sketch of S in Fig. 12.7 is not to scale).

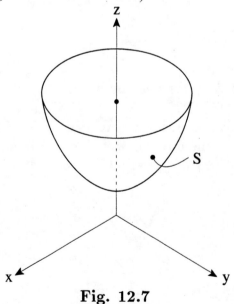

Fig. 12.7

■ **12.8.** S is the part of the spherical shell $x^2 + y^2 + z^2 = 4$ which lies in the first octant, with the normal pointing away from the origin. As a parameter representation of S we take

$$x = 2\cos u \sin v, \quad y = 2\sin u \sin v, \quad z = 2\cos v$$

(these are the formulas for spherical coordinates with $\rho = 2$). A point P on S has two kinds of coordinates: (i) (u,v) as a point on S; (ii) cartesian coordinates (x,y,z) as a

point in space. $\vec{n}$ is the normal of length 1 at $P = (u, v)$. The xyz-components of $\vec{n}$ are all functions of (u, v):
$$\vec{n} = (n_1(u,v), n_2(u,v), n_3(u,v)).$$

a) Find the three functions n_1, n_2, n_3. Hint: $\vec{n} = \frac{1}{r}\vec{r}$ because S is part of a sphere.

b) $\vec{F}$ is the vector field $\vec{F} = (1, z^2, 0)$. Find the normal component F_n of $\vec{F}$ at the following points: (i) $(u, v) = (0, \pi/4)$; (ii) (u, v), the "generic" point of S.

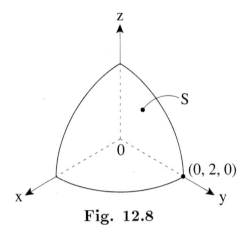

Fig. 12.8

Flux of a vector field across a surface - Riemann sums, estimates

■ **12.9.** S is the part of the cylinder $x^2 + y^2 = 4$ contained in the first octant between $z = 0$ and $z = 2$ shown in Fig. 12.9. The normal points away from the z-axis. $\vec{F}$ is the vector field $\vec{F} = (y + 1, x, z^2)$. Set up a Riemann sum of three terms to get the process of defining
$$\iint_S \vec{F} \cdot d\vec{A}$$
started. In particular: (i) Show the way you have subdivided S; (ii) show the points which you have picked in each of the three pieces; (iii) find the value of the Riemann sum.

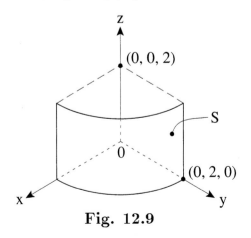

Fig. 12.9

■ **12.10.** S is the upper half $z \geq 0$ of the spherical shell of radius 3 centered at the origin (shown in Fig. 12.10), with the normal pointing outwards as shown, and $\vec{F}$ is the vector field $\vec{F} = (z+1)\vec{r}$ where $\vec{r} = (x, y, z)$. Give an estimate of $\iint_S \vec{F} \cdot d\vec{A}$, i.e., find numbers k and K such that $k \leq \iint_S \vec{F} \cdot d\vec{A} \leq K$.

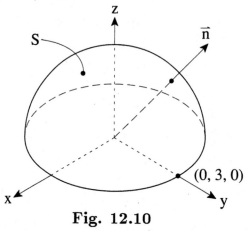

Fig. 12.10

■ **12.11.** S is the cylindrical surface shown on the left in Fig. 12.11. The top and bottom disks are part of S, and as normal of S we take the one pointing to the outside. The vector field $\vec{F}$ is defined by $\vec{F}(P) = (0, f(y), 0)$ where P is the point (x, y, z) and $f(y)$ the function shown below right.

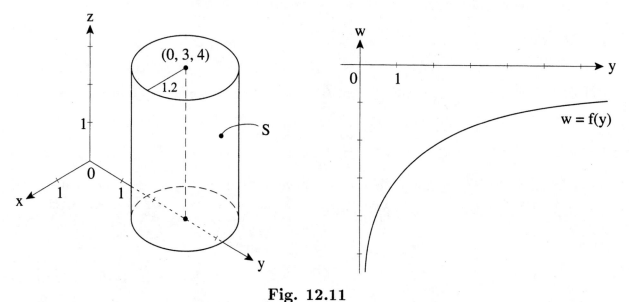

Fig. 12.11

a) In the system above left draw the field vector at the points $K(3, 4, 1)$ and $L(1.5, 1, 3)$.

b) Will the flux of $\vec{F}$ across S be positive? Or negative? Or zero? Or is it impossible to make any prediction? Hint: (1) What happens on the bottom disk? On the top? (2) A horizontal plane such as $z = 1$ intersects S in a circle C. Move a point on C and watch the normal component F_n of $\vec{F}$ on S.

Flux across a surface - Evaluation by inspection

■ **12.12.** S is the triangle with vertices $A(1,2,6)$, $B(3,0,0)$, $C(0,5,0)$ in Fig. 12.12. We take the normal whose third component is negative. Find the flux of the constant vector field $\vec{F} = (1,4,3)$ across S.

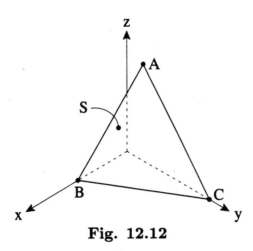

Fig. 12.12

■ **12.13.** S is a square with sides of length 3. One of its sides is on the z-axis. The normal of S is as shown in Fig. 12.13. The square turns around the z-axis, and its position is described by the angle θ. Find a value of θ for which the flux of the constant vector field $\vec{F} = (1, 2, 612)$ is (i) zero; (ii) lowest.

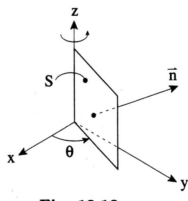

Fig. 12.13

■ **12.14.** The surface S consists of the 6 faces of a rectangular box of width 2, length 3 and height 4 (Fig. 12.14(i)). The normal of S points outward. $\vec{F}$ is the vector field $\vec{F} = (5, g(y), 0)$ where the graph of $g(y)$ is shown in Fig. 12.14(ii). Find the flux of $\vec{F}$ across S.

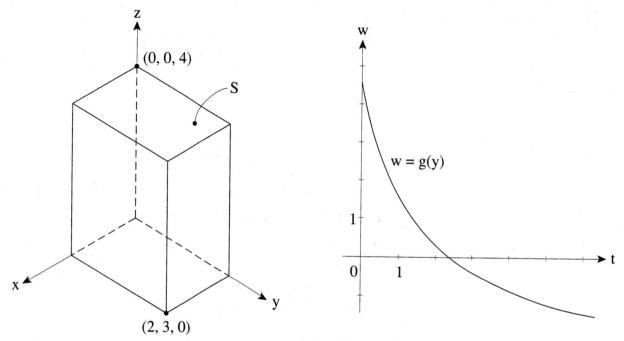

Fig. 12.14 (i) and (ii)

■ **12.15.** S is the spherical shell $x^2 + y^2 + z^2 = b^2$ with the outward normal, and $\vec{F} = (1/r^3)\vec{r}$ ($\vec{r} = (x, y, z)$, $r = |\vec{r}|$). Find the flux of $\vec{F}$ across S.

Flux across a surface - Evaluation by integration

■ **12.16.** S is the triangle ABC shown in Fig. 12.16, with the normal pointing away from the origin, and $\vec{F}$ is the vector field $\vec{F} = (y, 0, x)$. Find the flux of $\vec{F}$ across S.

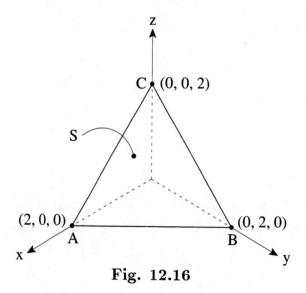

Fig. 12.16

■ **12.17.** Evaluate the integral of **12.9**.

■ **12.18.** C is the curve $z = x^2 - 6x + 8$ in the xz-plane, $0 \leq x \leq 2$, shown in Fig. 12.18. If we rotate C around the z-axis we obtain a surface S with parameter representation

$$x = r\cos\theta, \ y = r\sin\theta, \ z = r^2 - 6r + 8$$

(see **1.35**); note that $0 \leq r \leq 2$ and $0 \leq \theta \leq 2\pi$. We take the normal which points away from the z-axis. $\vec{F}$ is the vector field $\vec{F} = (x, y, z)$. Find the flux of $\vec{F}$ across S.

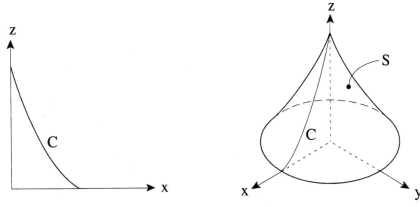

Fig. 12.18

■ **12.19.** $w = f(x, y, z) = x^2 + y^2 - z$. S is the level surface $w = 16$, with the normal pointing in the direction of $grad\, f$, and such that (x, y) lies in the triangle shown in Fig. 12.19. Find $\iint_S grad\, f \cdot d\vec{A}$.

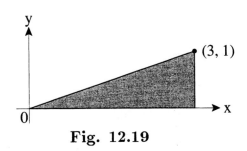

Fig. 12.19

■ **12.20.** S is the square OABC in the xy-plane shown in Fig. 12.20 (normal pointing up), and $\vec{F} = (y^2, 2, xy)$. To find the flux of $\vec{F}$ across S, the following reasoning is offered: "Since we are working in the xy-plane we apply the divergence form of Green's theorem (see **11**): The divergence of $\vec{F}$ is zero, and therefore the flux is zero." Is this reasoning correct? If yes, evaluate the flux from scratch (i.e., use integration) to confirm that the flux is zero. If no, where is the flaw, and what is the flux?

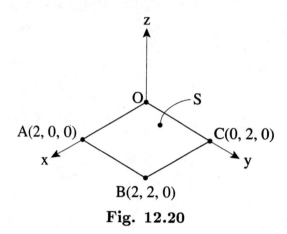

Fig. 12.20

■ **12.21.** S consists of the cylinder $x^2 + y^2 = 9$, $0 \le z \le 5$, and the top disc $x^2 + y^2 \le 9$, $z = 5$, i.e., S is a can with the bottom removed. We take the normal which points out of the can. $\vec{F}$ is the vector field $\vec{F} = (-y, x, x^2 z)$. Find the flux of $\vec{F}$ across S. Hint: What is the normal component of $\vec{F}$ on the vertical walls of the cylinder?

13 | The Theorem of Gauss (Divergence Theorem)

The boundary of solids, surfaces and curves

■ **13.1.** Below are given eight geometrical objects $M_1, \ldots, M_8$ by their formulas. For each $M = M_1, \ldots$ answer the following questions:

(i) Is M a curve, a surface or a solid?

(ii) The boundary ∂M of M might be empty. If it is not empty, do you expect it to be a surface? A solid? A pair of points? A curve?

(iii) Describe the boundary ∂M of M in words and by formulas.

(iv) Give the coordinates of a point U of M which is *not* on ∂M, and of a point V which is on ∂M.

$M_1 : x^2 + y^2 + z^2 \leq 9$; $M_2 : x^2 + y^2 + z^2 = 9$; $M_3 : x^2 + y^2 + z^2 = 9, z \geq 0$;
$M_4 : x^2 + y^2 + z^2 \leq 9, z \geq 0$; $M_5 : x^2 + y^2 \leq 4, z = 0$; $M_6 : x^2 + z^2 = 16, y = 4$;
$M_7 : x = 5\cos t, y = 5\sin t, z = 0, 0 \leq t \leq \pi$; $M_8 : x = 2 - u + 3v, y = 1 + 4u - v, z = 7u + 2v, 0 \leq u \leq 1, 0 \leq v \leq 1$.

■ **13.2.** K is the solid $1 \leq x^2 + y^2 + z^2 \leq 4$, i.e., the solid between two spherical shells of radius 1 and 2. Describe ∂K in words and formulas. Describe the exterior normal on the parts which make up ∂K.

■ **13.3.** S is the cylindrical surface $x^2 + y^2 = 4, 0 \leq z \leq 5$, shown in Fig. 13.3. T is the "top disk" $x^2 + y^2 \leq 4, z = 5$.

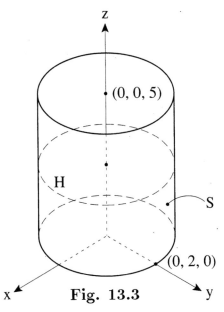

Fig. 13.3

a) Describe the boundary of the following figures $M_1, \ldots$: $M_1 = S$; M_2: S plus top disk T; M_3: solid cylinder $x^2 + y^2 \leq 4$, $0 \leq z \leq 5$; M_4: line segment connecting the points $(2, 0, 0)$ and $(2, 0, 5)$ on S.

b) H is the the circle $x^2 + y^2 = 4$, $z = 2.5$ (the circle on S halfway up). Describe two surfaces S_1 and S_2 which both have H as boundary.

The theorem of Gauss (the Divergence Theorem)

■ **13.4.** $\vec{F}$ is the constant vector field $\vec{F} = (3, 1, -7)$. S_1 is a torus (the surface obtained by rotating the circle $(x - 3)^2 + z^2 = 1$ around the z-axis, see Fig. 13.4 (i)), and the surface S_2 is the boundary of the potato-shaped solid shown in Fig. 13.4 (ii). We take the outward normal for both S_1 and S_2.

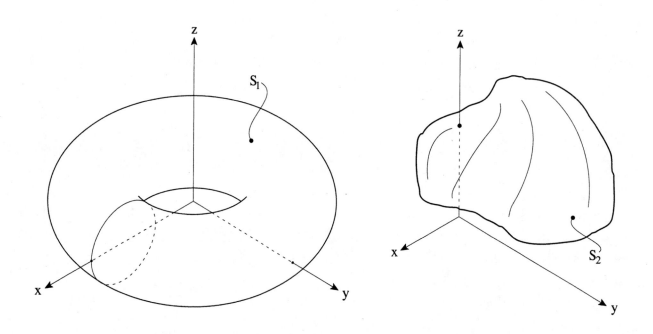

Fig. 13.4 (i) and (ii)

a) Find the flux of $\vec{F}$ across S_1.

b) Do you have enough information to find the flux of $\vec{F}$ across S_2? If yes, find the flux. If no, explain.

c) Give an example of a vector field $\vec{G}$ which is not constant, but for which you can find the flux across S_1 and S_2 by the reasoning used in a) and b).

■ **13.5.** $\vec{F}$ is the vector field of **12.16**, and we consider the surface S consisting of the three faces OAB, OAC, OBC of the tetrahedron $OABC$ of **12.16**. Use the answer to **12.16** to find the flux of $\vec{F}$ across S.

■ **13.6.** We go back to **12.11**. How would the divergence theorem work for the vector field $\vec{F}$ and the surface S of **12.11**? In a) and b) below you will be asked to set up, but not evaluate, the volume integral $\int\int\int div\,\vec{F}\,dV$ and the surface integral $\int\int \vec{F}\cdot d\vec{A}$. Both integrals will reduce to simpler integrals.

a) In the triple integral we take as dV a sliver of the solid cylinder that is parallel to the xz-plane, has height 4 and thickness dy. The triple integral reduces to an ordinary integral. Set up the latter.

b) For the surface integral we observe that the flux across S equals four times the flux of the *plane* vector field $\vec{G} = (0, f(y))$ across the *plane* curve $x^2 + (y-3)^2 = (1.2)^2$. The flux integral in the plane is a curvilinear integral. Set it up.

c) The divergence theorem says that the values of the integrals of a) and b) are the same. The flux across S is positive. The integral of a) is manifestly positive since $f(y)$ is increasing. The integral b) must be positive, too. What do you respond to your colleague who says "How can this integral be positive since $f(y)$ is negative all the time?"

■ **13.7.** Use the theorem of Gauss to work **12.14**.

■ **13.8.** The surface S consists of the cylinder $x^2 + y^2 = a^2$, the top disk $x^2 + y^2 \leq a^2$, $z = 6$, and the bottom disk $x^2 + y^2 \leq a^2$, $z = 0$, as shown on Fig. 13.8. $\vec{F}$ is the vector field defined by
$$\vec{F} = (y^2, x^2, z^2).$$

Find $\int\int_S \vec{F}\cdot d\vec{A}$.

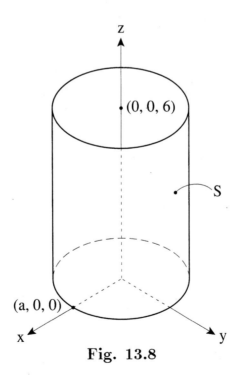

Fig. 13.8

■ **13.9.** We consider the two spherical shells $S_1 : x^2+y^2+z^2 = 4$ and $S_2 : x^2+y^2+z^2 = 9$, both equipped with their exterior normals. $\vec{F}$ is the vector field $\vec{F} = (1/r^3)\vec{r}$ of **12.15**.

  a) Find $div\ \vec{F}$.

  b) Can you apply the theorem of Gauss to find the flux of $\vec{F}$ across S_1 and across S_2? If yes, do it. If no, explain.

  c) Reconcile your answer to b) with your answer to **12.15**.

  d) Now K is the solid between S_1 and S_2 (Fig. 13.9 shows the lower half of K which can be thought of as a thickened shell). On S_1 we take the normal pointing away from the origin (shown at A), and on S_2 we take the normal pointing towards the origin (shown at B). Can you apply the divergence theorem to find the flux of $\vec{F}$ across ∂K? If yes, do it. If no, explain.

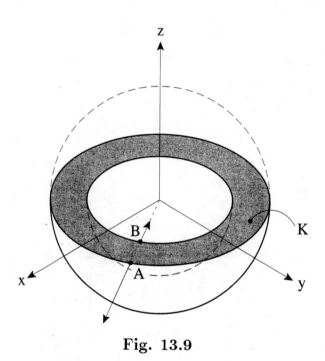

Fig. 13.9

■ **13.10.** (*This is a longer problem. It practices also pushing through many steps and dealing with a function whose formula is not given explicitly.*)

The function $w = g(t)$ of the single variable t is given by the graph shown in Fig. 13.10 a). We use $g(t)$ to define a vector field $\vec{F}$ by

$$\vec{F} = g(r^2)\vec{r} \quad (\vec{r} = (x, y, z),\ r = |\vec{r}|).$$

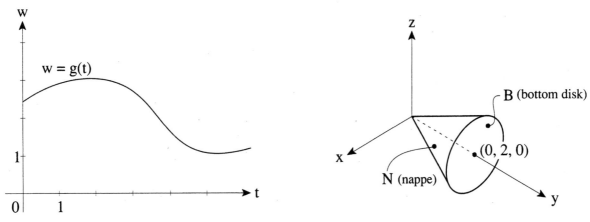

Fig. 13.10

a) Find the components of $\vec{F}$ at the point $P(1,1,1)$.

b) (i) Express $\text{div } \vec{F}$ in terms of r, g and the derivative g'. (ii) Evaluate $\text{div } \vec{F}$ at $P(1,1,1)$.

c) K is the solid circular cone of height 2 (Fig. 13.10 on the right). Its bottom disk B is given by $x^2 + z^2 \leq 1$, $y = 2$. Your answer to b) shows that it would be hard to evaluate $\iiint_K \text{div } \vec{F}\, dV$ as a triple integral. Use the divergence theorem to evaluate the same integral. Hints: (1) ∂K consists of B and the nappe N. What is the normal component F_n of $\vec{F}$ on N and B? (2) Evaluate the flux as integrals of F_n over N and B. (3) You will integrate F_n over B, and for this you will use polar coordinates in the plane $y = 2$. Call these coordinates ρ, θ in order not to get confused by the $\vec{r} = (x, y, z)$ which figures in F_n.

■ **13.11.** K is the straight circular solid cone shown in Fig. 13.11. Its boundary ∂K consists of the bottom disk B and the nappe N.

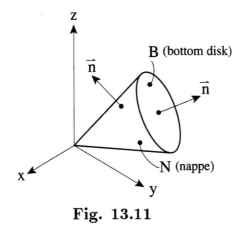

Fig. 13.11

$\vec{F}$ is the constant vector field $\vec{F} = (a, b, c)$. We have the following information on the flux of $\vec{F}$ across the nappe N: $\iint_N \vec{F} \cdot d\vec{A} = 3.75$. Do we have enough information to find the flux of $\vec{F}$ across the bottom disk B (normal as shown)? If yes, find the flux. If no, spell out what else is needed to be able to evaluate the flux across B.

■ **13.12.** S is the half-shell $x^2 + y^2 + z^2 = b^2$, $z \geq 0$, with the normal pointing away from the origin. $\vec{F}$ is the vector field defined by

$$\vec{F} = (3x + z^2, y, x^2 + y^2).$$

Find the flux of $\vec{F}$ across S. Hint: Look at the solid halfball in Fig. 13.12 of which S is part of the boundary, and apply the divergence theorem.

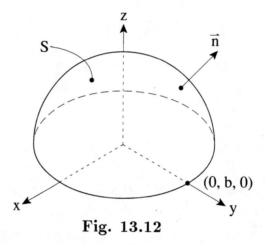

Fig. 13.12

■ **13.13.** In Fig. 13.13 D is a domain in the xy-plane. Its boundary ∂D in the xy-plane has a parameter representation

$$x = f(u), \quad y = g(u), \quad a \leq u \leq b.$$

K is the solid cylinder of height 1 above D.

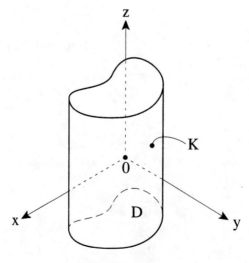

Fig. 13.13

$\overrightarrow{F}$ is the vector field $\overrightarrow{F} = (P(x,y), Q(x,y), 0)$ (note the absence of z in all three components). The divergence theorem says

$$\iint_{\partial K} \overrightarrow{F} \cdot d\overrightarrow{A} = \iiint_K \operatorname{div} \overrightarrow{F} \, dV.$$

a) Set up the surface integral on the left as far as you can; you will end up with a line integral in the xy-plane. Hints: $\partial K = $ top + bottom + vertical wall; use the parameter representation of C to get a parameter representation of the vertical wall.

b) Set up the triple integral on the right as far as you can; you will end up with a double integral in the xy-plane.

c) Use a) and b) to rewrite the statement of the divergence theorem above. You fall back onto a topic of an earlier chapter. Which one?

■ **13.14.** $\overrightarrow{F} = (-y^2, x^2, z^2)$.

a) K_1 is the solid ball $x^2 + y^2 + (z-2)^2 \leq 4$. Find the flux of $\operatorname{curl} \overrightarrow{F}$ across ∂K_1 (exterior normal). Note that we ask for the flux of $\operatorname{curl} \overrightarrow{F}$ and *not* for the flux of $\overrightarrow{F}$.

b) K_2 is the potato-shaped solid of **13.4 (ii)**. We take the exterior normal on its boundary ∂K_2. Do we have enough information to find the flux of $\overrightarrow{F}$ across ∂K_2? Do we have enough information to find the flux of $\operatorname{curl} \overrightarrow{F}$ across ∂K_2? If yes, find the flux.

c) Now $\overrightarrow{G}$ is the "general" vector field $\overrightarrow{G} = (a(x,y,z), b(x,y,z), c(x,y,z))$. Find $\operatorname{div} \operatorname{curl} \overrightarrow{G}$.

d) Your answer to c) is the general fact which underlies your answers to a) and b). Formulate the upshot of a), b) and c) in general: "The vector field $\overrightarrow{G}$ is defined for all points in space. Then the flux of .... across .... is ...."

14 | The Theorem of Stokes

■ **14.1.** For this chapter you have to be thoroughly comfortable with the notion of boundary. Therefore, do problem **13.1**. If you have already done so, do it again.

■ **14.2.** The circle C lies in the plane $\alpha: 3x - y + z = 5$. Its center is at $(x, y, z) = (2, 6, 5)$, and its radius is R. The orientation of C is induced by the normal $\vec{n}$ of α which has a positive third component. Figure 14.2 is not to scale.

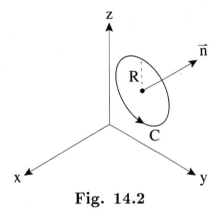

Fig. 14.2

The vector field $\vec{F}$ is defined by $\vec{F} = (0, 3z + y, 2y)$. Find the circulation of $\vec{F}$ along C.

■ **14.3.** In Fig. 14.3 S is the spherical halfshell $x^2 + y^2 + z^2 = b^2$, $z \geq 0$, with the normal pointing outwards.

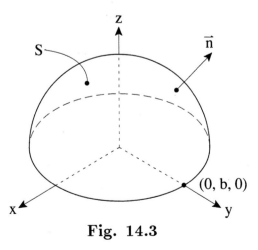

Fig. 14.3

The vector field $\vec{F}$ is defined by $\vec{F} = (z(1 + y), x + z, xyz)$. Find $\iint_S \operatorname{curl} \vec{F} \cdot d\vec{A}$.

■ **14.4.** $\vec{F} = (x^2 + y^2 + z^2)^{-1}(y, xz, xyz)$, and C is a circle of radius b in the xy-plane (see Fig. 14.4 (i) upper left).

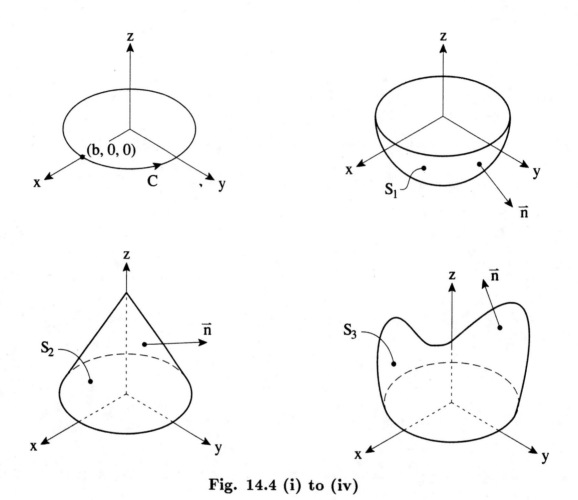

Fig. 14.4 (i) to (iv)

a) Find the circulation of $\vec{F}$ along C, with C oriented as shown.

b) S_1 is a spherical halfshell, S_2 is a circular cone, and S_3 is a two-horned surface (see above). All three surfaces have C as boundary, and we take normals pointing away from the z-axis. Find the flux of $curl\,\vec{F}$ across these three surfaces. Hint: Pay attention to the normals and the orientation of C.

c) There is a general fact underlying your answer to b). Complete the following statement: *Consider a vector field $\vec{G}$ and two surfaces S and S^*. Choose normals for both surfaces and orient their boundaries accordingly. If $\partial S = \partial S^*$ as oriented curves, then the flux of ... across ... and ... is ... .*

■ **14.5.** S_3 is the surface of the preceding problem, and $\vec{H} = (x^2 + y^2)(2x, 2y, 3z^2)$.

  a) Find the flux of $curl\ \vec{H}$ across S_3.

  b) Find the tangential component H_{tan} of $\vec{H}$ along ∂S_3.

  c) There is a general fact underlying your answer to a). Complete the statement: Consider a vector field $\vec{G}$ and a surface S. If G_{tan} is equal to ... on ..., then the flux ... .

  d) If $curl\ \vec{G}$ is zero on the surface S then the general fact which you just stated is a triviality. Go back to the vector field $\vec{H}$ of a) and find $curl\ \vec{H}$. Your result shows that the general fact of c) is *not* a triviality.

■ **14.6.** The surface T, in Fig. 14.6, is the torus of **13.4**. Let us consider a small circle C of radius b on T. We mutilate T by removing the interior of C from T so that we obtain a new surface T_1; T_1 is a torus with a hole. To indicate that the shape of the new surface T_1 depends on the radius b chosen, we call it $T_1(b)$.

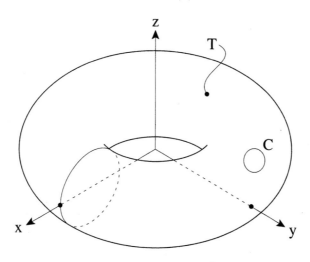

Fig. 14.6

  a) Describe the boundary ∂T_1 of the mutilated torus T_1. Orient the boundary consistently with the exterior normal of the torus.

Now we take a general vector field $\vec{F}$ which is defined on T. Stokes' theorem says:

$$\int_{\partial T_1(b)} \vec{F} \cdot d\vec{r} = \iint_{T_1(b)} curl\ \vec{F} \cdot d\vec{A}.$$

  b) Use your answer to a) to rewrite the left side of Stokes' theorem; write down the theorem with the new left side.

  c) Now we let b, the radius of the small circle which we cut out of the torus, tend to zero. Go to the statement of Stokes' theorem in b): What is the limit of the left side? What is the limit of the right side?

  d) Nowhere in a) to c) did we use the fact that the surface T is a torus. We could have taken any *closed* surface S (by a *closed* surface in space is meant the boundary of a solid in space). Formulate your answer to c) as a general fact: *S is a closed surface and $\vec{F}$ is a vector field defined on S. Then the flux of ... across ... is equal to ....*

e) Let S and $\vec{F}$ be as in d). We apply Stokes' theorem directly (i.e., without mutilating the surface, etc.) and obtain of course $\int_{\partial S} \vec{F} \cdot d\vec{r} = \int\!\int_S \text{curl}\, \vec{F} \cdot d\vec{A}$. By d) both integrals should be zero. How can we see this directly? Hint: Look at the left side. S is a closed surface - what is its boundary?

Note. The general fact formulated in **13.14** d) is yet another confirmation of what you found in d) of the preceding problem.

■ **14.7.** $\vec{F}$ is the vector field $\vec{F} = (3x + 1, \frac{-z}{y^2+z^2}, \frac{y}{y^2+z^2})$.

a) (i) Find $\text{curl}\,\vec{F}$. (ii) Describe the points where $\vec{F}$ is defined and where not.

b) C_1 is the circle $y^2 + z^2 = 1$, $x = 0$, with the orientation as shown on the left in Fig. 14.7. Find $\int_{C_1} \vec{F} \cdot d\vec{r}$.

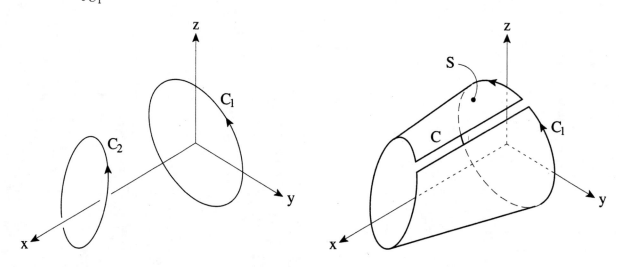

Fig. 14.7

C_2 is another circle which winds around the x-axis (see above left). We construct a surface S as follows: We wrap a cylinder around C_1 and C_2 and cut it up along the segment C (see above right). We take the outward normal for S.

c) The boundary ∂S of S is shown above right; it consists of several parts. Orient all parts consistently with the orientation of C_1 and mark the orientations with arrows.

Now we apply Stokes' theorem: $\int\!\int_S \text{curl}\,\vec{F} \cdot d\vec{A} = \int_{\partial S} \vec{F} \cdot d\vec{r}$.

d) Evaluate the left side of the preceding statement of Stokes' theorem.

e) Write out the right side as sum of integrals over the different parts of ∂S (we consider the cut C to have width zero, i.e. in integrating we go back and forth over the same segment C). What is finally the circulation of $\vec{F}$ along the tilted circle C_2?

f) Now we consider another circle C_3 which is not shown above. It lies in the plane $4x + 7y + 5z = 160$; its radius is 3 units, its center is at $(10, 10, 10)$, and its orientation is consistent with the normal $\vec{n} = (4, 7, 5)$ of the plane. Find the circulation of $\vec{F}$ along C_3. Hint: Look at the disk D whose boundary is C_3.

g) In this part, we discuss the general fact behind the process used in e) and f). – We consider a vector field $\vec{G} = (G_1(x, y, z), G_2(x, y, z), G_3(x, y, z))$ whose *curl* is zero. **If** it is possible to find a surface S such that

(i) $\partial S = C$ (ii) G is defined at all points of S

then the circulation of $\vec{G}$ along C vanishes as a consequence of Stokes' theorem.

Can we apply this kind of reasoning to our $\vec{F}$ and C_2 of above? If yes, do it and compare your result to your answer to e). If no, explain.

■ **14.8.** The vector field $\vec{F}$ is defined by

$$\vec{F} = (0, f(x), 0)$$

where $f(x)$ is a function of one variable. The surface S is a rectangle in the xy-plane with the normal pointing up, as shown in Fig. 14.8.

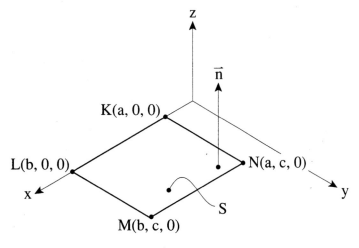

Fig. 14.8

Stokes' theorem says

$$\int_{\partial S} \vec{F} \cdot d\vec{r} = \iint_S \operatorname{curl} \vec{F} \cdot d\vec{A}.$$

Evaluate both sides as far as is possible. You fall back on an important formula of elementary calculus. Which one?

15 | Gauss' and Stokes' Theorem, Miscellanea

Problems on both theorems

■ **15.1.** In Fig. 15.1 K is the solid halfball $x^2 + y^2 + z^2 \leq 1$, $z \geq 0$; S is the halfshell $x^2 + y^2 + z^2 = 1$, $z \geq 0$; D is the disk $x^2 + y^2 \leq 1$, $z = 0$, and C is the circle $x^2 + y^2 = 1$, $z = 0$. $\vec{G}$ stands for the vector fields $(i)\ \vec{F} = (-3y, z, y)$; $(ii)\ \text{curl}\ \vec{F}$.

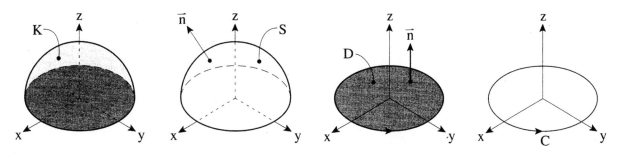

Fig. 15.1

Find the flux of $\vec{G}$ across: (1) ∂K; (2) S; (3) D, and find the circulation of $\vec{G}$ along C (total of 8 questions).

■ **15.2.** K, S and C are the halfball, halfshell and circle of the preceding problem. $\vec{F}$ is a vector field defined on K, and f is a scalar function defined on K. Consider the six integrals (1),...,(6):

(1) $\iint_{\partial K} \text{grad}\, f \cdot d\vec{A}$ (2) $\iiint_K \text{div}\, \vec{F}\, dV$ (3) $\iint_S f\, dA$

(4) $\iiint_K f\, dV$ (5) $\int_C \vec{F} \cdot d\vec{r}$ (6) $\iint_S \text{div}\, \vec{F}\, dA$

Below are six other integrals (A),...,(F). Next to each of them write the number of those integrals (1),...,(6) equal to it no matter what $\vec{F}$ and f are. If none of (1),...,(6) corresponds write *none*:

(A) $\iint_{\partial K} \vec{F} \cdot d\vec{A}$ (B) $\iint_S \text{curl}\, \vec{F} \cdot d\vec{A}$ (C) $\int_C \text{grad}\, f \cdot d\vec{r}$

(D) $\int_C \text{curl}\, \vec{F} \cdot d\vec{r}$ (E) $\iiint_K \text{div}\, \text{grad}\, f\, dV$ (F) $\iint_{\partial K} \text{grad}\, \text{div}\, \vec{F} \cdot d\vec{A}$

Invariant definition of divergence and curl

■ **15.3.** S is the spherical shell $(x-a)^2 + (y-b)^2 + (z-c)^2 = 10^{-6}$ with the normal pointing outwards, as shown in Fig. 15.3.

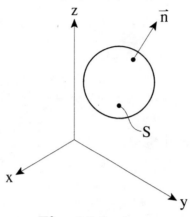

Fig. 15.3

We have the following information on a vector field $\vec{F}$:

$$\operatorname{div} \vec{F} = -7 \quad \text{at the point } (a, b, c).$$

Give the best estimate you can for the flux of $\vec{F}$ across S.

■ **15.4.** We have the following information on a vector field $\vec{F}$:

$$\operatorname{curl} \vec{F} = (1, 1, 4) \quad \text{at the origin.}$$

We are interested in the circulation of $\vec{F}$ over circles C of radius $b = 0.004$ centered at the origin, as shown in Fig. 15.4. Each of these circles lies in a plane α which passes through the origin.

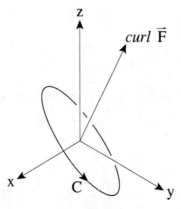

Fig. 15.4

Give the best estimate you can for the plane α (i.e., for its equation $ax + by + cz + d = 0$) so that the absolute value of the circulation $\int_C \vec{F} \cdot d\vec{r}$ is as close to zero as possible.

16 | Coordinate Transformations

Note to instructor. We consider a change of coordinates (e.g., from x, y to u, v) and curvilinear coordinates (e.g., spherical coordinates ρ, θ, ϕ) consistently as a map between planes (from the uv-plane to the xy-plane) or spaces (from $\rho\theta\phi$-space to xyz-space). This is the way polar, cylindrical and spherical coordinates were treated in Chapter 6; it is spelled out in **16.1** and assumed from that point on. We write O for the origin $(0,0)$.

Linear transformations

■ **16.1.** A and B are two points in the xy-plane as shown on the right in Fig. 16.1. All through this problem we will work with these two points which are kept fixed.

We use A and B to introduce new coordinates (u, v) in the xy-plane as follows: The uv-coordinates of a generic point P of the xy-plane are obtained by expressing $\overrightarrow{OP}$ as a linear combination of the vectors $\overrightarrow{OA}$ and $\overrightarrow{OB}$ (see below right):

$$\overrightarrow{OP} = u\,\overrightarrow{OA} + v\,\overrightarrow{OB}.$$

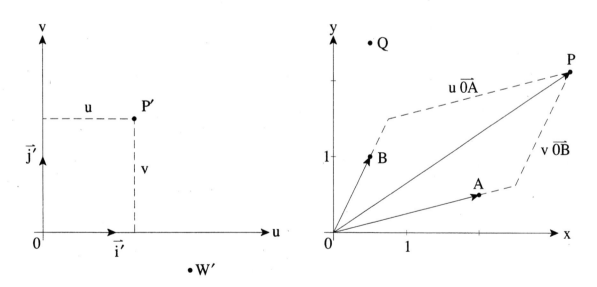

Fig. 16.1

For a given point P the scalars u, v are uniquely determined. Thus one and the same point P in the xy-plane has *two* sets of coordinates, viz. (x, y) and (u, v).

a) What are the uv-coordinates of A, B, O? Use ruler and pencil to find the uv-coordinates of the points P and Q shown above right.

b) Draw the point S in the xy-plane whose uv-coordinates are $(\frac{1}{2}, 2)$.

16—Coordinate Transformations

We interpret this new system of uv-coordinates in the xy-plane as a map $\mathbf{F}$ which assigns to a point P' in a uv-plane a point P in the xy-plane in the following way:

Look at the unit vectors $\vec{i'}$ and $\vec{j'}$ of the uv-plane (above left); a point P' in the uv-plane has cartesian coordinates (u, v) defined by $\overrightarrow{O'P'} = u\vec{i'} + v\vec{j'}$.

Now the map $\mathbf{F}$ assigns to the point P' in the uv-plane the point P in the xy-plane given by $\overrightarrow{OP} = u\overrightarrow{OA} + v\overrightarrow{OB}$.

In other words: If P' has cartesian coordinates (u, v) in the uv-plane, then $\mathbf{F}(P')$ is the point in the xy-plane whose uv-coordinates are (u, v) in the sense explained at the beginning of the problem.

c) (i) Draw the points A', B', Q' in the uv-plane whose corresponding points in the xy-plane are A, B, Q (i.e., $\mathbf{F}(A') = A$, $\mathbf{F}(B') = B$, ...). (ii) Draw the point W in the xy-plane which corresponds to the point W' shown in the uv-plane.

d) The x- and y-coordinates of $\mathbf{F}(P')$ are functions of (u, v), and similarly the uv-coordinates of a point in the xy-plane are functions of (x, y):

$$x = f(u, v) \qquad u = h(x, y)$$
$$y = g(u, v) \qquad v = k(x, y)$$

Find the four functions f, g, h, k. – Hint: Read off the xy-coordinates (a_1, a_2) of A and (b_1, b_2) of B. Then $(x, y) = u(a_1, a_2) + v(b_1, b_2)$, and from this you get $x = \ldots$ and $y = \ldots$. To obtain the functions h and k you solve the latter two equations for u and v.

The two functions f and g define the mapping $\mathbf{F}: (u, v) \mapsto (x, y)$ from the uv-plane to the xy-plane, and the functions h and k define the the inverse mapping $\mathbf{F}^{-1}: (x, y) \mapsto (u, v)$ from the xy-plane to the uv-plane. In summary, looking at the new coordinates in the xy-plane (as done at the beginning) and looking at the two maps $\mathbf{F}$ and $\mathbf{F}^{-1}$ amounts to the same.

■ **16.2.** a) Find a mapping $x = f(u,v)$, $y = g(u,v)$ from the uv-plane to the xy-plane such that the square $O'A'B'C'$ in the uv-plane corresponds to the parallelogram $OABC$ in the xy-plane (i.e., $O' \to O$, $A' \to A$ etc).

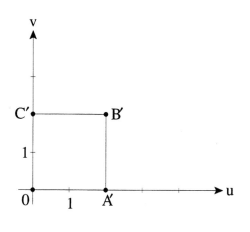
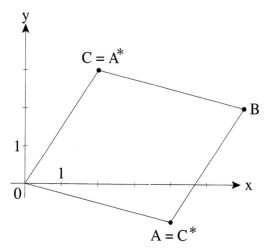

Fig. 16.2

b) Do the same for the parallelogram OA^*BC^* (i.e., $O' \to O$, $A' \to A^*$ etc).

■ **16.3.** Find a mapping $\mathbf{F}: (u,v) \mapsto (x,y)$ so that the square $O'A'B'C'$ corresponds to the tilted square $OABC$ in Fig. 16.3. Note that both squares have the same edge of length 1. Draw the figure E which corresponds to the triangle E'.

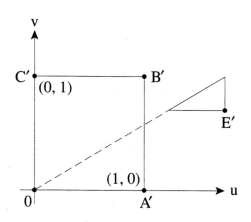
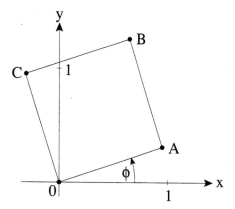

Fig. 16.3

■ **16.4.** Let C be the parabola in the xy-plane obtained by rotating the standard parabola $y = x^2$ by the angle $\phi = \frac{\pi}{6}$ (see Fig. 16.4).

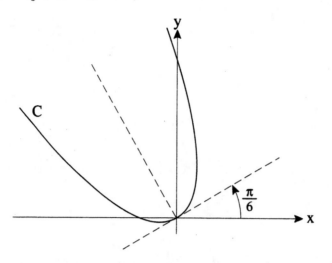

Fig. 16.4

a) Use the preceding problem to define a mapping $\mathbf{F}$ so that the equation of the corresponding curve C' in the uv-plane is as simple as possible.

b) Use a) to write down the equation of C in terms of x and y.

■ **16.5.** Find a mapping $x = f(u,v,w)$, $y = g(u,v,w)$, $z = h(u,v,w)$ from uvw-space to xyz-space such that the unit cube $O'A'B'C'\ldots$ in uvw-space corresponds to the parallelepiped $OABC\ldots$ in xyz-space (Fig. 16.5 is not to scale).

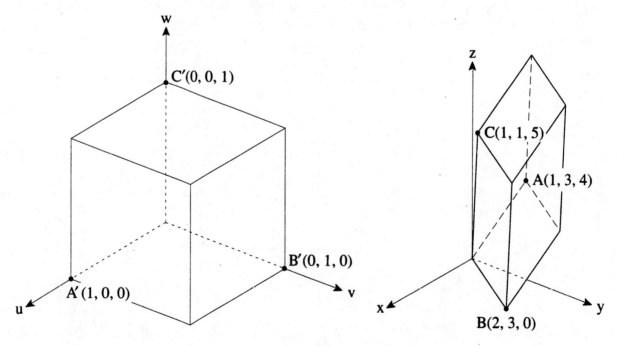

Fig. 16.5

The Jacobian of linear transformations

■ **16.6.** We consider the map $\mathbf{F}: (u, v) \mapsto (x, y)$ defined by

$$x = f(u, v) = u + 3v,$$
$$y = g(u, v) = 2u - v.$$

The vectors $\overrightarrow{OA}$ and $\overrightarrow{OB}$ in the xy-plane which give rise to $\mathbf{F}$ (in the way discussed in **16.1**) are shown on the right in Fig. 16.6.

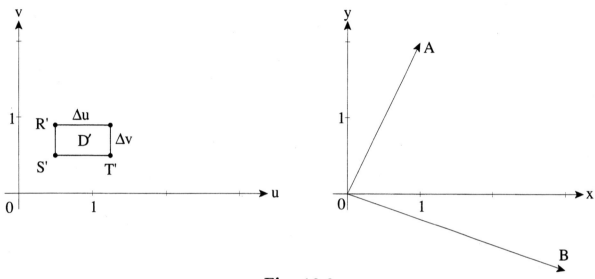

Fig. 16.6

a) D' is a small rectangle in the uv-plane whose edges of length Δu and Δv are parallel to the uv-axes (see above left). Draw the figure D in the xy-plane which corresponds to D'.

b) (i) What is the surface area $A(D')$ of D'?
(ii) Express the area $A(D)$ of D in terms of Δu and Δv (use vector algebra to find the area of a parallelogram spanned by two vectors).
(iii) E is the parallelogram in the xy-plane spanned by $\overrightarrow{OA}$ and $\overrightarrow{OB}$. Find the corresponding figure E' in the uv-plane.
(iv) Use (ii) and (iii) to find $A(E)$.

c) Let G' be any figure of area $A(G')$ in the uv-plane (e.g., an ellipse and its interior), and let G be the corresponding figure in the xy-plane. What kind of relation would you expect between $A(G')$ and $A(G)$? Why?

It is highly plausible that the answer to c) is $A(G) = k \cdot A(G')$ where $k = 7$. The constant factor k was obtained as the area of a certain parallelogram.

d) Find the Jacobian matrix $\frac{\partial(f,g)}{\partial(u,v)}$ where $f(u, v)$ and $g(u, v)$ are the functions which define the map $\mathbf{F}$, and whose formulas are given at the beginning. Find the Jacobian determinant (the determinant of the Jacobian matrix).

You realize that the absolute value of the Jacobian determinant equals the constant factor k mentioned just before d). Is this a coincidence, or is there a general pattern? Explain.

■ **16.7.** We return to the map **F** of **16.3**.

   a) Find its Jacobian determinant.

   b) Let E' be any figure of area $A(E')$ in the uv-plane, and E its corresponding figure in the xy-plane. What will be $A(E)$? Substantiate your answer in two ways - by using a formula, and by looking at the geometric way the map **F** works.

■ **16.8.** Define a map $\mathbf{F}: (u,v) \mapsto (x,y)$ whose Jacobian determinant equals -2. Show in an xy-system the vectors $\overrightarrow{OA}$ and $\overrightarrow{OB}$ which define **F**.

■ **16.9.** Consider the map $\mathbf{F}: (u,v,w) \mapsto (x,y,z)$ of **16.5**.

   a) Find the volume of the parallelepiped spanned by the three vectors $\overrightarrow{OA}$, $\overrightarrow{OB}$ and $\overrightarrow{OC}$.

   b) Find the Jacobian determinant of the map **F**.

   c) Rewrite **16.6** from c) on, adapted to the present map $\mathbf{F}: (u,v,w) \mapsto (x,y,z)$, and work the rewritten questions c) and d).

Curvilinear coordinates

■ **16.10.** We consider polar coordinates in the xy-plane

$$x = r \cos \theta = f(r, \theta), \quad y = r \sin \theta = g(r, \theta)$$

at the point $P'_0 : (r, \theta) = (2, \frac{\pi}{3})$. P_0 is the corresponding point in the xy-plane.

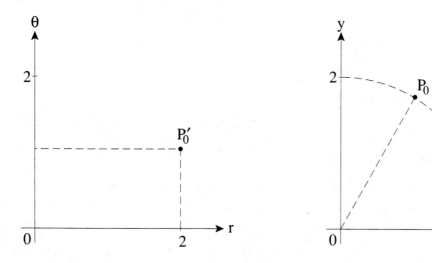

Fig. 16.10

a) $f(r, \theta)$ is a function of the two variables r and θ. As such it has a linear approximation at P'_0, and we call it $f_1(r, \theta)$. Similarly, $g_1(r, \theta)$ is the linear approximation of $g(r, \theta)$ at the same point P'_0. Write out the formulas for f_1 and g_1.

Now we have *two* maps from the uv-plane to the xy-plane on our hands:

the map $\mathbf{F}: (r, \theta) \mapsto (f(r, \theta), g(r, \theta))$ which defines polar coordinates;

the map $\mathbf{F}_1 : (r, \theta) \mapsto (f_1(r, \theta), g_1(r, \theta))$ where f_1, g_1 are the linear approximations of f, g at P_0'.

We call $\mathbf{F}_1$ the linear approximation of $\mathbf{F}$ at P_0'.

b) D' is the rectangle in the $r\theta$-plane shown in Fig. 16.10 b)(i).

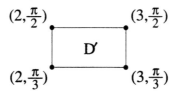

Fig. 16.10 b) (i)

The map $\mathbf{F}$ transforms D' into a figure D in the xy-plane, and the map $\mathbf{F}_1$ transforms D' into a figure D_1. Draw D and D_1 into the big xy-system in Fig. 16.10 b)(ii). Colloquially one could call the figure D_1 "the linear approximation of the figure D."

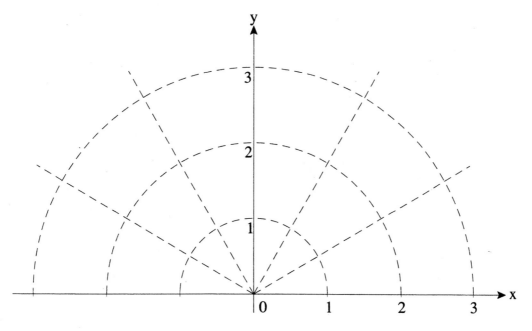

Fig. 16.10 b) (ii)

c) Measure the area $A(D')$ of D' and $A(D_1)$ of D_1. Define the number k by

$$A(D_1) = k\, A(D'),$$

i.e., k is the factor by which areas are multiplied under the linear map $\mathbf{F}_1$.

d) Find the numerical value of the Jacobian determinant of $\mathbf{F}_1$ at P_0' and compare it with k of c). What should be the answer to d) according to theory?

■ **16.11.** In this problem, the general situation underlying **16.10** will be worked out. We consider curvilinear coordinates (u,v) in the xy-plane, i.e., a map $\mathbf{F}:(u,v) \mapsto (x,y) = (f(u,v), g(u,v))$, together with a fixed point P_0' in the uv-plane and its corresponding point P_0 in the xy-plane:

$$P_0' : (u,v) = (u_0, v_0); \quad P_0 = (f(u_0, v_0), g(u_0, v_0)) = (x_0, y_0)$$

(see Fig. 16.11). The linear approximation of $\mathbf{F}$ at P_0' will be called $\mathbf{F}_1$:

$$\mathbf{F}_1 : (x,y) \mapsto (f_1(u,v), g_1(u,v))$$

(here f_1, g_1 are the linear approximations of f, g at P_0'). In addition, we deal with a rectangle D' in the uv-plane whose edges of length Δu and Δv are parallel to the uv-axes. In the drawing below, the figure D (with corners P_0, Q, R, S) corresponds to D' under the map $\mathbf{F}$, and the parallelogram D_1 (with corners P_0, K, L, M) corresponds to D' under the linear approximation $\mathbf{F}_1$.

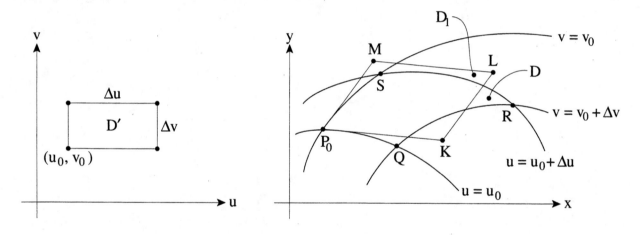

Fig. 16.11

a) Express the xy-components of $\overrightarrow{P_0 K}$ and $\overrightarrow{P_0 M}$ in terms of the partial derivatives of f and g (evaluated at P_0'), i.e., find numbers $a_1, \ldots$ such that

$$\overrightarrow{P_0 K} = a_1 \vec{\imath} + a_2 \vec{\jmath}, \quad \overrightarrow{P_0 M} = b_1 \vec{\imath} + b_2 \vec{\jmath}.$$

Hint: $\overrightarrow{P_0 K} = \overrightarrow{OK} - \overrightarrow{OP_0}$; find K' in D' and find the components of $\overrightarrow{OK}$ as the xy-coordinates of K.

b) (i) Find the area $A(D_1)$ of D_1 in terms of the partial derivatives of f and g. Define k by $A(D') = k\, A(D_1)$, and write out a formula for k. (ii) Write out the Jacobian determinant of $\mathbf{F}$ at P_0', and compare it with k. (iii) Complete the statement: *For small Δu and Δv, the area of D, the figure in the xy-plane corresponding to the rectangle D' under the map $\mathbf{F}$, is approximately* ....

■ **16.12.** We consider curvilinear coordinates (u, v) defined by

$$x = f(u,v) = e^u \sin v, \quad y = g(u,v) = e^u \cos v.$$

Figure 16.12 shows parts of curves $u = constant$ (circles) and $v = constant$ (rays) in the xy-plane.

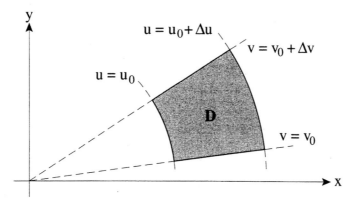

Fig. 16.12

a) Find values u_0 and v_0 such that for small Δu and Δv the area of the shaded figure D is approximately $c \cdot \Delta u \Delta v$ where c has the following values (three questions):

$$(i)\ c = 1; \quad (ii)\ c = 26; \quad (iii)\ c = 0.0084.$$

b) D' is the figure in the uv-plane which corresponds to the figure D in the xy-plane under the map $x = \ldots, y = \ldots$ above. Set up a uv-system with units of 2 cm and draw the figures D' corresponding to the three questions of a).

■ **16.13.** Some of the coordinate curves $u = const$ and $v = const$ of curvilinear coordinates

$$x = f(u,v), y = g(u,v)$$

are shown in Fig. 16.13. In answering the questions you should assume that the coordinate curves which are not shown look as the sketch suggests (i.e., there are no unexpected irregularities).

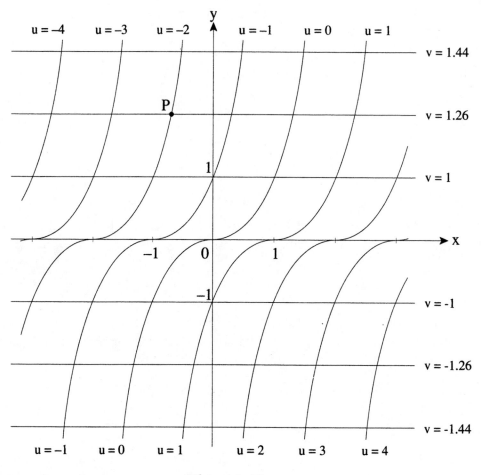

Fig. 16.13

a) C is the curve in the xy-plane whose equation is $u - v = 1$. Draw C as well as you can into the the xy-system above.

b) Would you expect the Jacobian determinant at P (i.e., at $(u, v) = (-2, 1.26)$) to be positive? Negative? Greater than 1? Less than 1? Why? Write down reasons for your answer in such a form that your colleague will understand them.

c) Find a point where the Jacobian determinant is likely to be zero. Write down reasons for your answer in such a form that your colleague will understand them.

d) Which (if any) of the following pairs of functions $(f(u,v), g(u,v))$ could be the curvilinear coordinates under discussion:

(1) (u, v^3) (2) (v, u^3) (3) $(u, (u+v)^3)$ (4) $(u+v, v^3)$ (5) $(u+v, u^3)$ (6) $(v^3, u+v)$.

■ **16.14.** We consider curvilinear coordinates (u, w) in the xz-plane (note the w in (u, w) and the z in xz):

$$x = f(u, w), \; z = g(u, w),$$

but we will be using them only in the halfplane $x \geq 0$. Consider the uw-plane as part of uvw-space and the xz-plane as part of xyz-space. We construct curvilinear coordinates (u, v, w) in xyz-space by assigning to (u, v, w) the point (x, y, z) obtained by rotating the point $(f(u, w), 0, g(u, w))$ around the z-axis by the angle v, as shown in Fig. 16.14.

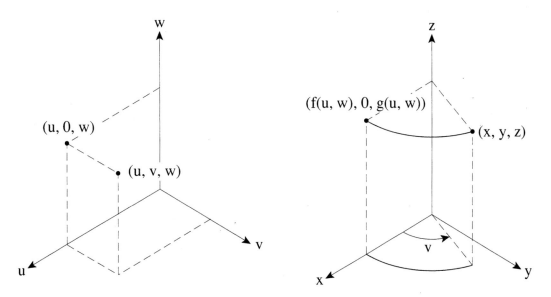

Fig. 16.14

Then x, y, z are given by the following formulas:

$$x = f(u, w) \cos v,$$
$$y = f(u, w) \sin v,$$
$$z = g(u, w),$$

where v will vary only between 0 and 2π.

a) Describe each of the following three coordinate surfaces of the uvw-system we just constructed:

(i) $u = 3$, (ii) $v = \pi/6$, (iii) $w = 2$.

The description will be in terms of the coordinate curves of the uw-coordinates in the xz-plane.

b) We have *two* Jacobian determinants on our hands: The one from the (u,w)-system in the xz-plane, and the one from the (u,v,w)-system in xyz-space. Compute the latter and express it in terms of the former.

c) Take for $f(u,w)$ and $g(u,w)$ the functions $f(u,w) = u$ and $g(u,w) = w$. Write out the formulas for the uvw-system. You fall back on well known curvilinear coordinates. Which? Reconcile your answer to b) with the situation on hand.

d) Now we take $f(u,w) = u \sin w$ and $g(u,w) = u \cos w$. That is, (u,w) are polar coordinates in the xz-plane except that the angle w is counted from the z-axis down (and not from the x-axis up as usual). Write out the formulas for the uvw-system. Again you fall back on well known curvilinear coordinates. Which? Reconcile your answer to b) with the situation on hand.

Change of variables in multiple integrals - Linear transformations

■ **16.15.** D is the parallelogram shown in Fig. 16.15. Evaluate $\iint_D (2x - y) dA$.

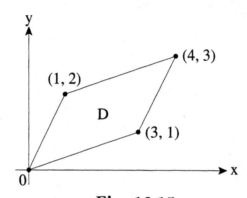

Fig. 16.15

■ **16.16.** We go back to **16.3**. There we considered the mapping

$$x = \cos\phi \cdot u - \sin\phi \cdot v, \quad y = \sin\phi \cdot u + \cos\phi \cdot v.$$

Under this mapping the unit square $O'A'B'C'$ in the uv-plane corresponds to a tilted square $OABC$ in the xy-plane. The Jacobian of this mapping is equal to 1.

Now we let $f(p,q)$ be a function of two variables. Consider the following statement ($OABC$ and $O'A'B'C'$ are still the squares of **16.3**):

"The Jacobian is equal to 1. Therefore

$$\iint_{OABC} f(x,y) dx dy = \iint_{O'A'B'C'} f(u,v) du dv$$

no matter what the function f is."

Is the statement true? Whatever your answer is, give reasons.

■ **16.17.** We introduce new coordinates in the xy-plane by

$$x = 2u + v, \quad y = u + 2v.$$

a) K is the parallelogram in the xy-plane bounded by the curves $u = 1$, $u = 3$, $v = 0$, $v = 2$ in the xy-plane. Into the systems in Fig. 16.17 draw K and the corresponding K' in the uv-plane. Then set up $\iint_K x\,dx\,dy$ as an integral in the uv-plane.

b) M' is the triangle in the uv-plane with vertices $(u,v) = (0,0), (2,-1), (1,-2)$. Into the systems below draw M' and the corresponding M in the xy-plane. Then set up $\iint_{M'}(3u + 5)\,du\,dv$ as an integral in the xy-plane.

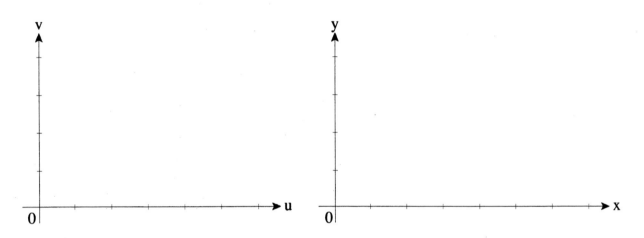

Fig. 16.17

■ **16.18.** K is the parallelepiped in xyz-space of **16.5**. Use your answer to **16.9** to set up the integral $\iiint_K z\,dx\,dy\,dz$.

■ **16.19.** We introduce new coordinates (u, v, w) in xyz-space by the formulas

$$x = u + w$$
$$y = 2u + v$$
$$z = u - v + w.$$

a) M is a solid in xyz-space. Its volume is 56.42 units. Find the volume of the corresponding solid M' in uvw-space.

b) Find the determinant of the Jacobian matrix $\frac{\partial(u,v,w)}{\partial(x,y,z)}$.

c) T is the solid tetrahedron in xyz-space with vertices $A(1,2,1)$, $B(0,1,-1)$, $C(1,0,1)$, $D(0,0,0)$, and $f(x,y,z) = x - z$. Set up $\iiint_T f(x,y,z)\,dx\,dy\,dz$ as an integral in uvw-space.

16—Coordinate Transformations

Change of variables in multiple integrals - Nonlinear transformations

■ **16.20.** The curvilinear coordinates (u, v) in the xy-plane are defined by

$$x = u + v, \quad y = u^2 - v^2 + 2.$$

a) Find the equations of the coordinate curves $u = constant$ and $v = constant$. What is the shape of these curves?

b) R' is a small rectangle in the uv-plane, with sides parallel to the axes. It follows from your answer to a) that the corresponding domain R in the xy-plane is bounded by four pieces of parabolas. Use the Jacobian at $(u, v) = (1, 1)$ to estimate the surface area of R (the drawing of R in Fig. 16.20 is rough and not to scale).

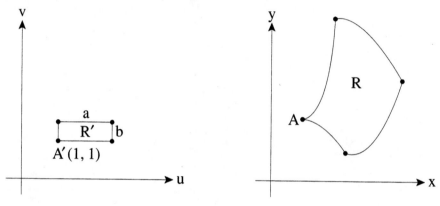

Fig. 16.20

c) Find the area of R by integration.

d) Take $a = 0.1$ and $b = 0.04$, and compare the area of R with the estimate you gave in b). What is the percentage error of the estimate?

■ **16.21.** The surface S is given by the parameter representation

$$x = f(u, v), \quad y = g(u, v), \quad z = h(u, v).$$

We can use the first two functions to define curvilinear coordinates (u, v) in the xy-plane:

$$x = f(u, v), \quad y = g(u, v).$$

The mapping which defines these uv-coordinates can be thought of as the mapping which defines S followed by the projection onto the xy-plane:

$$(u, v) \mapsto (f(u, v), g(u, v), h(u, v)) \mapsto (f(u, v), g(u, v)).$$

D' is a rectangle in the uv-plane, D_S is the corresponding patch on the surface, and D is the figure in the xy-plane which corresponds to D'.

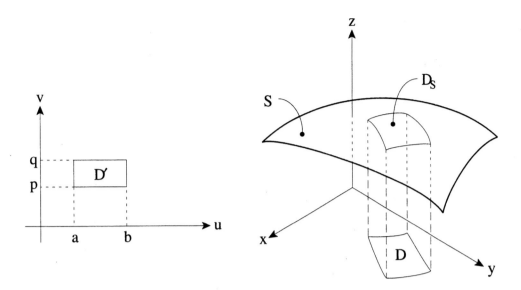

Fig. 16.21

a) Set up $\iint_D dA$ as an integral in the uv-plane.

b) Set up the integral which gives the surface area of the patch D_S.

c) To answer a) you used the formula for the change of variables in multiple integrals, and for b) you used the procedure of evaluating surface integrals. Look carefully at your answer to b) and explain how it is related to your answer to a).

■ **16.22.** The function $f(t)$ of one variable t is given by its graph in Fig. 16.22.

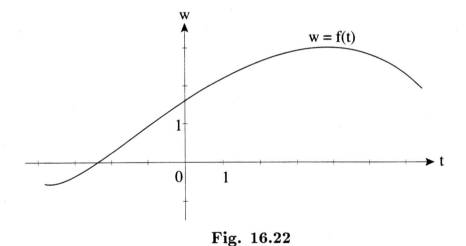

Fig. 16.22

We define curvilinear coordinates (u, v) in the xy-plane by

$$x = f(u), \quad y = v.$$

a) Find the figure D' in the uv-plane which corresponds to the rectangle D shown in Fig. 16.22 b).

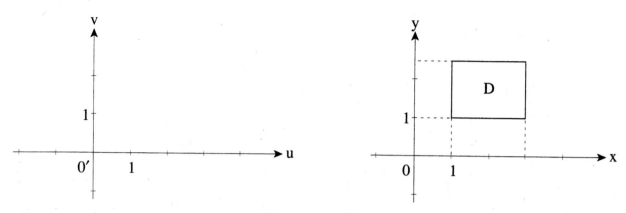

Fig. 16.22 b)

b) $H(x)$ is yet another function of one variable. We can always consider it as a function of two variables: $(x, y) \mapsto H(x)$. In both (i) and (ii) below push the computations as far as you can. The D in question is the one discussed in a).

(i) Set up $\iint_D H(x)\,dx\,dy$ as an integral in the xy-plane.

(ii) Set up the same integral as an integral in the uv-plane.

(iii) The fact that the integral of (i) equals the integral of (ii) expresses a formula of elementary calculus. Which one?

■ **16.23.** The formulas $x = u^2 - v^2$, $y = u + v + w$, $z = w$ define curvilinear coordinates in xyz-space. K is the solid in xyz-space which corresponds to the solid K' in uvw-space. The sketch of K in Fig. 16.23 is very rough.

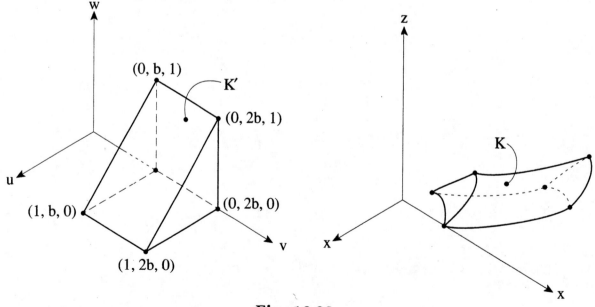

Fig. 16.23

Find: (i) $\iiint_K dx\,dy\,dz$; (ii) $\iiint_{K'} du\,dv\,dw$.

■ **16.24.** K is the solid torus obtained by rotating the disc $(x-A)^2 + z^2 \le b^2$, $y=0$ around the z-axis (see Fig. 16.24 left).

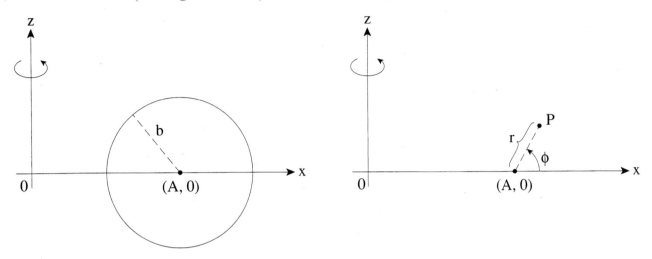

Fig. 16.24

In order to evaluate a triple integral over K it is expedient to find curvilinear coordinates (u,v,w) such that K is easily expressed in them. We do this using the method of **16.14** as follows. If P is a point in the xz-plane we describe its location by the distance r from $(A,0,0)$ and the angle ϕ, as shown above right. The formulas are

$$x = A + r\cos\phi, \quad z = r\sin\phi.$$

a) Rotate the xz-plane around the z-axis by the angle θ following the method of **16.14**. You obtain curvilinear coordinates (r,θ,ϕ) (careful–they are *not* the spherical coordinates). Write down the formulas which express each of x, y, z in terms of r, θ, ϕ.

b) The solid torus K in xyz-space corresponds to a solid K' in $r\theta\phi$-space. Describe K'.

c) Use **16.14** to find the Jacobian determinant for our r, θ, ϕ.

d) Set up the integral $\iiint_K z^2 \, dx\,dy\,dz$. Note. In the integral the Jacobian determinant J has to appear in absolute value. Use the fact $A > b$ to convince yourself that J is always positive (i.e., you do not have to worry about the absolute value).

Answers

1 | Lines, Curves, Planes and Surfaces

- **A–1.1.** m_1: $(1+4t, 1+2t)$, m_2: $(1+3t, 3-3t)$, m_3: $(1+5t, 3-3t)$.
- **A–1.2.** a) see Fig. A–1.2; b) $(3.26)\sqrt{13}$; c) $2x+3y-2=0$.

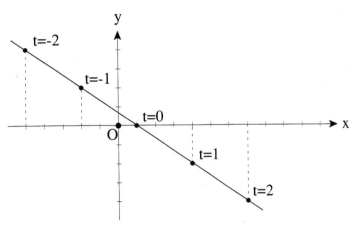

Fig. A–1.2

- **A–1.3.** a) (c_1+9, c_2-12) and (c_1-9, c_2+12); b) 3.26.
- **A–1.4.** a) $(2t, 3+5t)$; b) $(t, -2t+4)$; c) $(2-(t/2), t)$.
- **A–1.5.** a) see Fig. A–1.5; b) 0.6 sec.

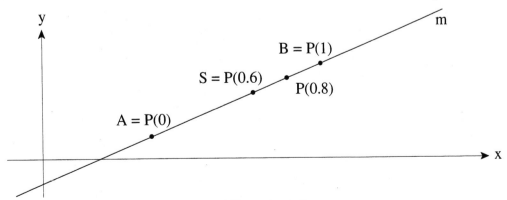

Fig. A–1.5

- **A–1.6.** a)(i) $c = -\frac{1}{5}$, (ii) $\overrightarrow{OR} = \overrightarrow{OA} + \frac{4}{5}\overrightarrow{AB}$; b) $\overrightarrow{OP} = \overrightarrow{OA} + (\frac{4}{5} - \frac{u}{5})\overrightarrow{AB}$, $t = \frac{4}{5} - \frac{u}{5}$.

■ **A–1.7.** a) see Fig. A–1.7; b) incr.; c) const.; d) $x = (t^3/5) \cdot 4/\sqrt{17}$, $y = 5 + (t^3/5) \cdot (-1)/\sqrt{17}$.

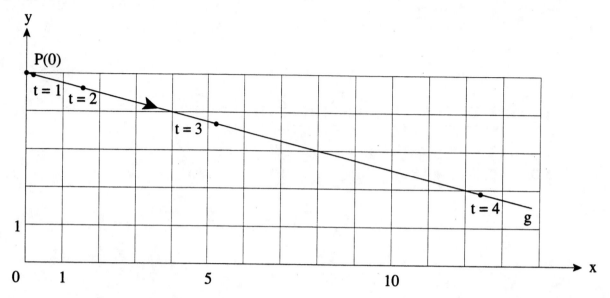

Fig. A–1.7

■ **A–1.8.** a) see Fig. A–1.8; b) no.

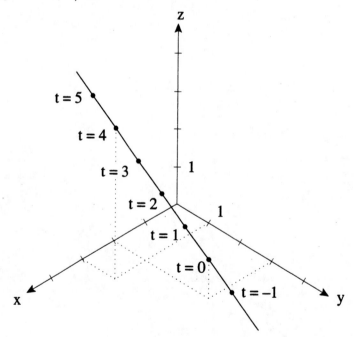

Fig. A–1.8

■ **A–1.9.** a) $\vec{r} = (1, 0, 10) + t \cdot (-4, -2, 10)$; b) see Fig. A–1.9; c) $(2.5) \cdot 2\sqrt{30}$, 2.5.

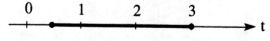

Fig. A–1.9

Answers: Ch 1   137

■ **A–1.10.** $(-7+5t, -3+4t, -4+2t)$.

■ **A–1.11.** a) $(4-3t, 4-4t, 5-3t)$; b) $\sqrt{34}$; c) $t\sqrt{34}$; d) $t=5/3$, $(-1, -8/3, 0)$; e) write $\vec{w}=(1/\sqrt{34})(3,4,3)$, and then the representation is $(4,4,5)+u\vec{w}$.

■ **A–1.12.** $(-2,-9,0)$, $(0,-5,2)$.

■ **A–1.13.** C_1: (i) $(2+t, 16+28t)$, (ii) $(2-28t, 16+t)$; C_2: (i) $(6+3t, 28+38t)$, (ii) $(6-38t, 28+3t)$; C_3:(i) $(1-5t, 2+18t)$, (ii) $(1+18t, 2+5t)$.

■ **A–1.14.** a) see Fig. A–1.14; b) $a'(6)=0.52$, $b'(6)=0.83$, length $|(a'(6), b'(6))|$ is 0.95 and should be 1.

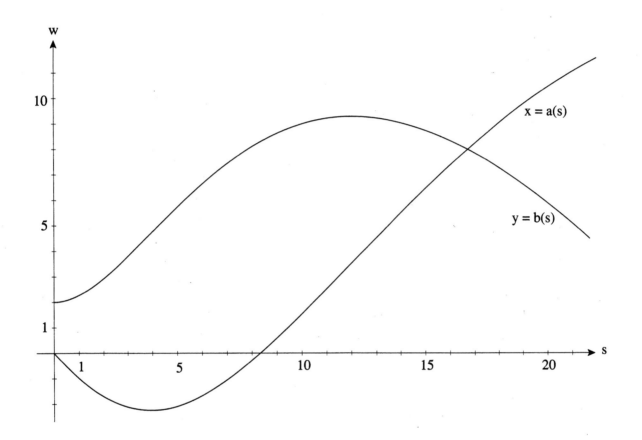

Fig. A–1.14

■ **A–1.15.** a) $(t, \cosh t)$; b) $\sinh t$; c) $(\sinh^{-1} s, \sqrt{1+s^2})$.

■ **A–1.16.** a) $(3.9, 2.9, 2.4)$; b) $(0.5, 0.78, -1.25)$, 1.6; c) 0.16.

■ **A–1.17.** a) $(3\cos t, 1, 3\sin t)$; b) see Fig. A–1.17.

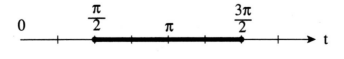

Fig. A–1.17

■ **A–1.18.** a) 0.090449632; b) 0.09; c) 0.09045.

■ **A–1.19.** (i) $(8, 3, 1)$; (ii) $(2.75, 0, 2.125)$; (iii) $(8+14u, 3+2u, 1-3u)$; (iv) $14x+2y-3z = 115$.

■ **A–1.20.** We write A for $2\theta + \frac{\pi}{2}$. Then: a) $(A\cos\theta, A\sin\theta)$; b) $\int_0^{3\pi/2} \sqrt{A^2+4}\, d\theta$.

■ **A–1.21.** a) $(x,y,z) = (r\frac{\sqrt{2}}{2}, r\frac{\sqrt{2}}{2}, f(r))$; b) $(x,y,z) = (\sqrt{2}+t\frac{\sqrt{2}}{2}, \sqrt{2}+t\frac{\sqrt{2}}{2}, f(2)+tf'(2))$.

■ **A–1.22.** a) $(x,y,z) = (\theta\cos\theta, \theta\sin\theta, \theta)$; b) $x + 2\pi y + z = 4\pi$.

■ **A–1.23.** a) see Fig. A–1.23; b) 3; c) $2x + y + 2z = 6$.

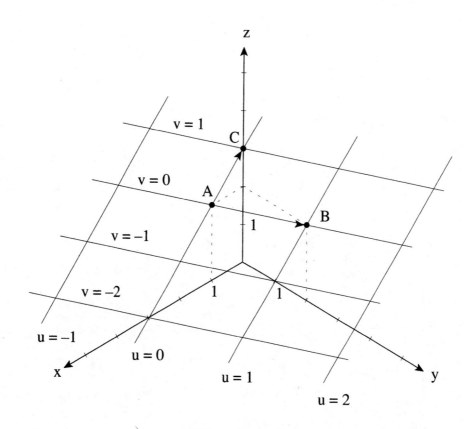

Fig. A–1.23

■ **A–1.24.** $x = x$, $y = y$, $z = (2/3)x + y$.

■ **A–1.25.** a) $(-1, 5, 2)$; b) $(2+v, 2+v, -1-2v)$; c) no, yes, yes.

■ **A–1.26.** a) $\overrightarrow{OP} = \overrightarrow{OA} + u\overrightarrow{AB} + v\overrightarrow{AC} = (4-5u, 2+u-2v, 3u+5v)$; b) triangle with vertices $(u,v) = (0,0), (1,0)$ and $(0,1)$.

■ A-1.27.

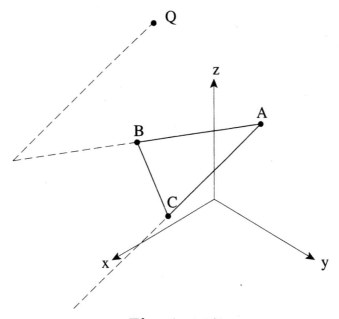

Fig. A-1.27

■ A-1.28. a) $(-1,1)$; b) $(u,v) = (0,1)$, $(p,q) = (1,0)$; c) (i) $p = v - u/2$, $q = u/2$; (ii) $u = 2q$, $v = p + q$.

■ A-1.29. $\overrightarrow{OP} = (7,6,4) + u(1,1,2) + v(4,2,1)$

■ A-1.30. $(2,2,2) + t(-32, 28, 1)$

■ A-1.31. a) see Fig. A-1.31; b) $(4+t, 18+8t, 4)$; c) with (p,q) as the parameters in the tangent plane we have $x = 4 + p$, $y = 18 + 8p + q$, $z = 4 + 4q$; d) $(1+3t, (2+4t) + (1+3t)^2, (2+4t)^2)$; e) $\int_0^1 \sqrt{365 + 1384t + 1348t^2}\, dt$; f) shaded patch in Fig. A-1.31 below; g) $9\Delta u \Delta v$.

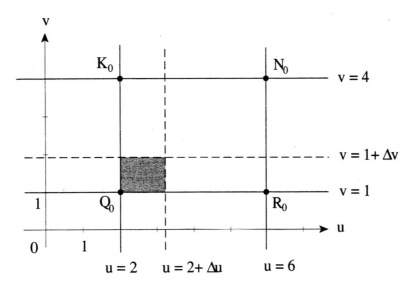

Fig. A-1.31

■ A–1.32. a) $x=x$, $y=y$, $z=xy$; b) $(3,y,3y)$ and $(x,5,5x)$, both straight lines; b) S_1: sphere, S_2: cylinder.

■ A–1.33. S_1: (i) $2x+4y+z=15$, (ii) $(1+2t, 2+4t, 5+t)$; S_2: (i) $4x+5y+z=10$, (ii) $(1+4t, 1+5t, 1+t)$; S_3: (i) $-y+2z=2$, (ii) $(2,-t,1+2t)$.

■ A–1.34. a) Consider a point P on the ray which is r units away from the z-axis. At time t its x- and y-coordinates are $x=r\cos\theta$, $y=r\sin\theta$, and its z-coordinate is $z=3\theta$; b) $(x,y,z)=(\frac{1}{2}r, \frac{\sqrt{3}}{2}r, \pi)$, a straight line; c) $3\sqrt{3}\,x - 3y + 4z = 4\pi$.

■ A–1.35. a) $(x,y,z)=(r\cos\theta, r\sin\theta, f(r))$; b) $r=c$: circle in a plane $z=const$; $\theta=c$: a copy of the curve $z=f(x)$ in the halfplane $\theta=c$; c) $(rf'(r)\cos\theta, rf'(r)\sin\theta, -r)$.

■ A–1.36. a) $(x,y,z)=(r\cos\theta, r\sin\theta, b-\frac{b}{a}r)$; b) rectangle with vertices $(r,\theta)=(\frac{a}{2},0)$, $(a,0)$, $(a,2\pi)$, $(\frac{a}{2},2\pi)$; c) The tangent plane contains g and is perpendicular to the xz-plane. Its equation is $\frac{x}{a}+\frac{z}{b}=1$. d) $(-b,0,-a)$; e) same plane as in c): $x=u$, $y=v$, $z=b-\frac{b}{a}u$;

f) the idea is to find a point on the cone such that $x=y=z$: $x=y=z=\frac{ab}{a+b\sqrt{2}}$.

■ A–1.37. First representation: $(x,y,z)=(x,y,x^2+y^2)$ (i.e., the xy-coordinates of a point on S are its parameters); (i) $x^2+y^2 \leq k^2$; (ii) $(2x, 2y, -1)$; second representation (rotating $z=x^2$ around the z-axis and using **1.35**): $(x,y,z)=(r\cos\theta, r\sin\theta, r^2)$; (i) rectangle with vertices $(r,\theta)=(0,0), (k,0), (k,2\pi), (0,2\pi)$; (ii) $(-2r^2\cos\theta, -2r^2\sin\theta, r)$.

2 | Functions of Two or More Variables

■ **A–2.1.** See Fig. A–2.1.

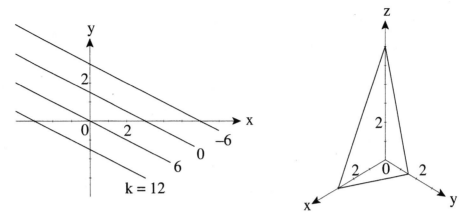

Fig. A–2.1 a) and b)

■ **A–2.2.** $f(x,y) = \frac{2}{7}x - \frac{4}{7}y + \frac{27}{7}$

■ **A–2.3.** See Fig. A–2.3.

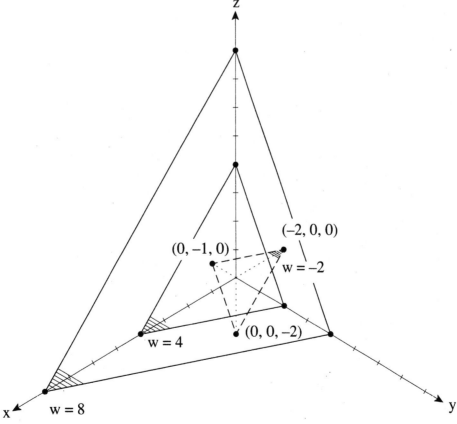

Fig. A–2.3

■ **A–2.4.** a) $2x + y + z = 8$. b) $p(x,y) = -2x - y + 8$, a function of two variables. c) $q(x,y,z) = 2x + y + z$, level surface $q = 8$.

■ **A–2.5.** P_1 and Q_1 are the projections of P and Q onto the xy-plane. a) See Fig. A–2.5 a) left; b) see Fig. A–2.5 b) right.

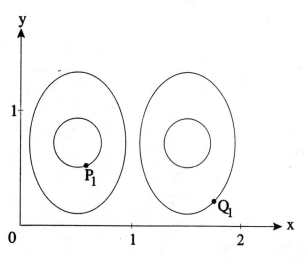

Fig. A–2.5 a) and b)

■ **A–2.6.** a), b), c) see Fig. A–2.6; d) $t = 70.8$; e) $x \notin [213.3, 215.8]$.

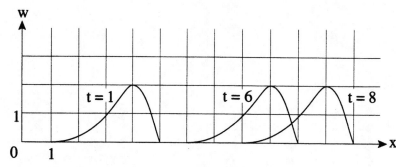

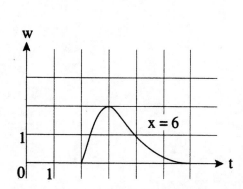
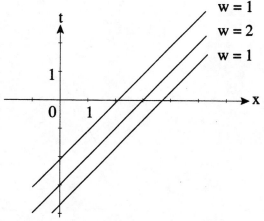

Fig. A–2.6 a) and b) and c)

■ **A–2.7.** Circles $x^2 + y^2 = 4$ and $x^2 + y^2 = 10$.

■ **A–2.8.** $P(2,0)$ and $Q(2,3)$; there are many correct answers.

■ **A–2.9.** a) See Fig. A–2.9; b) less.

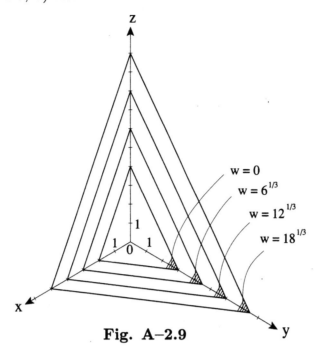

Fig. A–2.9

■ **A–2.10.** a) $R(\frac{\pi}{6},0,0)$, $S(0,\frac{\pi}{6},0)$; b) The level surface consists of infinitely many spheres $x^2+y^2+z^2 = (\sqrt{74}+2\pi k)^2$ and $x^2+y^2+z^2 = ((2k+1)\pi - \sqrt{74})^2$ where k is any integer; c) spheres $x^2+y^2+z^2 = (k\pi)^2$ where k is any integer.

■ **A–2.11.** a) and b) the planes $x=0$, $y=0$ and $z=0$; c) S: $z = \frac{1}{xy}$, hyperbolas $y = \frac{1}{x}$ and $y = \frac{1}{433x}$ in the xy-plane; d) the function $F(x,y,z) = xyz$ of *three* variables has level *surfaces*, and each of these level surfaces has level *curves*. In c) the problem was to draw two level curves of one of the level surfaces of $F(x,y,z)$.

■ **A–2.12.** A level set of a function of one variable is either empty or consists of points on the t-axis. $g(t) = -t^2$: $t=1$ and $t=-1$, $t=0$, empty; $g(t) = \sin t$: $t = \frac{3\pi}{2} + 2k\pi$ where k is any positive or negative integer, empty.

■ **A–2.13.** a) 5 parameters $p_1,\ldots,p_5$: $x_1 = 3p_1 - 2p_2 + 5p_3 - p_4 + 3p_5 + 15$, $x_2 = p_1$, $x_3 = p_2,\ldots, x_6 = p_5$; b) one parameter t: $x = t$, $y = 5$. "What is the advantage of such a parameter representation? It provides easy control over all points of the level set $z = 4$ as follows. For *any* choice of t we obtain a point of $z = 4$, and any point of $z = 4$ defines uniquely a parameter value t. In other words, the points (x,y) of the level set $z = 4$ are in 1:1 correspondence with the points t of the 1-dimensional t-axis."

3 | Partial Derivatives, the Chain Rule

■ **A–3.1.** a) First positive, then negative. b) See Fig. A–3.1.

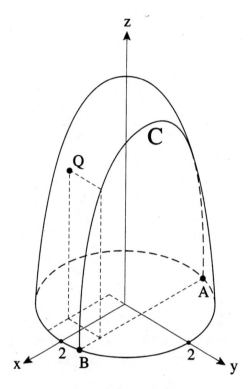

Fig. A–3.1 b)

■ **A–3.2.** a) (i) -4; (ii) 5.012. b) and c) see Fig. A–3.2.

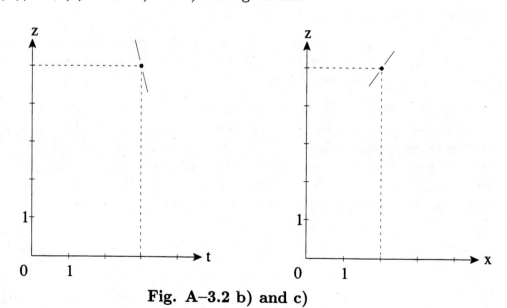

Fig. A–3.2 b) and c)

■ **A–3.3.** a) $f_x = -2.5$, $f_y = 0.8$. b) $Q(3, 2)$.

■ **A–3.4.** -0.58

■ **A–3.5.** $(a^2 + ab + b^2) \cdot G''(ax + by + c)$

■ **A–3.6.** a) $p_{xx} + (u+v)p_{xy} + uv\, p_{yy} + p_y$;
b) $(-\cos x/\cos y)$, $(1/\cos^2 y)(\cos y \sin x + \cos^2 x \tan y)$.

■ **A–3.7.** There are many answers, for example $Q : t = -2$, i.e., $Q(-5, 0)$, $R : t = 1$, i.e. $R(4, 3)$.

■ **A–3.8.** $(u, v) = (-\frac{2}{5}, -\frac{8}{5})$

■ **A–3.9.** We write $p(t)$ for the function $M(a(t), b(t), c(t), d(t))$. Then the answer is $p'(1) = 11$.

4 | Linearization

■ **A–4.1.** a) See Fig. A–4.1. b) -1.4

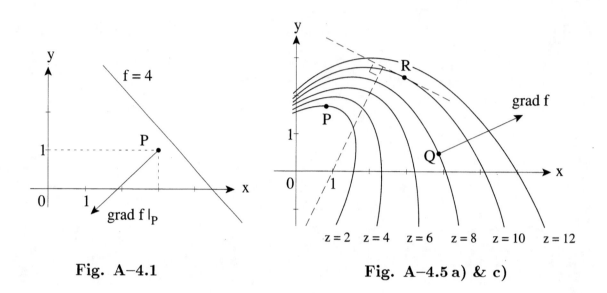

Fig. A–4.1   Fig. A–4.5 a) & c)

■ **A–4.2.** $Q(2, -1, 5)$

■ **A–4.3.** a) (i) = (ii): $(2A + B)/\sqrt{5}$; (iii): Yes. $f(x,y)$ is a *linear* function. Its formula is so simple that the average rate of change of f between R and S (computed in (i)) is the same as the directional derivative computed in (ii), no matter what the points R and S are. b) Yes, the linear functions $g(t) = Mt + N$. They have the property that their average rate of change between any two t-values t_0 and t_1 is equal to their constant derivative. For the graphs of these functions it means that the slope of the secant between any two points is equal to the slope of the tangent of the graph. This is obvious because the tangent of a straight line is the line itself.

■ **A–4.4.** $\frac{4}{3}$

■ **A–4.5.** a) and c) see Fig. A–4.5 a) & c) above right; b) less than; near Q the level curves are further apart than near P, and this indicates that near Q the value of $f(x, y)$ changes less per horizontal distance than near P. In other words, the largest directional derivative at Q will be less than the largest directional derivative at P.

■ **A–4.6.** a) −2.7; b) see Fig. A–4.6; c) 10.8

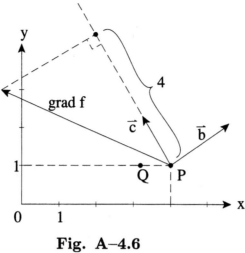

Fig. A–4.6

■ **A–4.7.** first negative, then positive
■ **A–4.8.** See Fig. A–4.8.

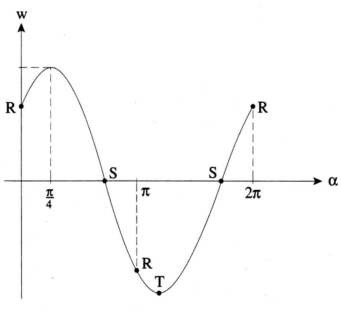

Fig. A–4.8

■ **A–4.9.** $2r + s + 8t = 0$; $(0, 8, -1)$ and $(1, -2, 0)$.
■ **A–4.10.** $g(Q) \approx 16 \pm b\frac{12}{\sqrt{5}}$, $g(R) \approx 16 \pm b\sqrt{48}$.

■ A–4.11. a) See Fig. A–4.11; b) $\operatorname{grad} g = F'(r)\frac{1}{r}(x,y,z)$; $r = \sqrt{14}$ for R and $3\sqrt{14}$ for S. First positive, then negative.

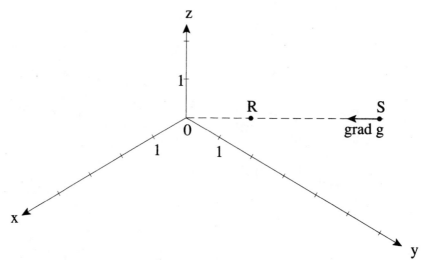

Fig. A–4.11

■ A–4.12. $g_{lin;3}(t) = \sqrt{10} + (0.3)\sqrt{10}\,(t - 3)$

■ A–4.13. a) and b): $f_{lin;P_0} = -16 + 12x + 3y = 8 + 12(x - 1) + 3(y - 4)$; c) see below.

(1): 8	(2): (12,3)	(3): 8	(4): (12,3)
(5): 8.14865663	(6): (11.98331857, 3.07883752)	(7): 8.15	(8): (12,3)
(9): 58.09475019	(10): (29.04737510, 17.42842506)	(11): 35	(12): (12,3)

d) We write S for the surface $z = f(x,y) = (xy)^{\frac{3}{2}}$.
(i) and (ii): $f_{lin;P_0}(x,y) = f(x_0,y_0) + \operatorname{grad} f|_{P_0} \cdot (x - x_0, y - y_0)$. Set $(x,y) = (x_0,y_0)$ in this formula and you obtain (i). Take the partial derivatives with respect to x and y of the formula and you obtain (ii).
(iii): Above points in the xy-plane close to $(1,4,0)$, the vertical distance between the tangent plane at $(1,4,8)$ and the surface S is small. In other words: (5) is close to (7).
(iv): The linear approximation is a linear function, and any such function has a constant gradient. This explains (4)=(8)=(12). Since $f(x,y)$ is not linear $\operatorname{grad} f$ is not constant, and this explains that (2), (6) and (10) are different from each other.
(v) The point $(3,5,0)$ is sufficiently far away from $(1,4,0)$ so that above $(3,5,0)$ the vertical distance between the surface S and the tangent plan at $(1,4,8)$ has become large.

■ A–4.14. $G(2,3) = -17$, $\operatorname{grad} G = (1,-8)$ at $(2,3)$.

■ A–4.15. a) $k(p_0 \Delta V + V_0 \Delta p + \Delta p \Delta V)$; b) $\Delta T \approx k(p_0 \Delta V + V_0 \Delta p)$; the graph of the linear approximation $T_{lin}(p,V)$ of $T = kpV$ at (p_0, V_0) is a plane in pVT-space. c) The difference is $k \Delta p \Delta V$; it is the vertical distance between the surface $T = kpV$ and the plane $T = T_{lin}(p,V)$ above the point $(p_0 + \Delta p, V_0 + \Delta V, 0)$.

■A–4.16. a) $F_1 = 4 + 2.8(x+1) + 1.3(y-1)$; b) see Fig. A–4.16.

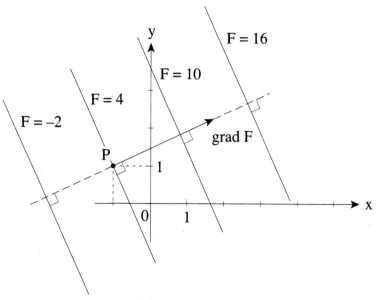

Fig. A–4.16

■A–4.17. At $(3, -1, 5)$: $H = 2.5$ and $grad\,H = (7, 0, -2)$; nothing can be said for the point $(7, 0, -2)$.

■A–4.18. a) $g_1 = 4 + 2(x-1) - 2(y-1) + 1(z-1) + 5(w-1)$; b) $\frac{16}{\sqrt{26}}$; c) take for example the direction of $-grad\,g$; d) $S: 2xzw^3 - y^2zw = 1$, $S_1: 2x - 2y + z + 5w = 6$; we need 3 parameters p, q, r:

$$x = p$$
$$y = q$$
$$z = -2p + 2q - 5r + 6$$
$$w = r.$$

■A–4.19. Point $(0, 0, 1)$, $6 + 7 \cdot 10^{-427}$

■A–4.20. a) $x = 0.1995$, $f(x) = 0.398003745$; b) $(\frac{\pi}{2} + 0.0017, 0.0054)$.

5 | Optimization

■**A–5.1.** a) local max at $t = 1.2, 7.3$; local min $t = 5.2, 8.8$; b) (i) local max at $t = 1.2, 7.3, 11$, local min at $t = 0, 5.2, 8.8$; (ii) global max at $t = 11$, global min at $t = 5.2$; c) (i) and (ii) local max and global max at $t = 2$, local min and global min at $t = 4$; d) $[0, 6]$; e) $[6, 8]$.

■**A–5.2.** a) and b): Local minima at $\pm(2.8, -0.7)$; saddle point at $(0, 0)$; somewhere around $\pm(4.5, -3)$ there are critical points of unascertainable nature; c) triangle: minimum at $(2.8, -0.7)$, maximum at $(1.4, -4)$; rectangle: minimum at D (or G?), maximum at F; ellipse: minimum at $(-1.8, -4.3)$, maximum at $(-4.3, -5.1)$.

■**A–5.3.** a) (i) 1.6; (ii) none; (iii) $x_1 = \pm 1.1$ and ± 5.1; b) Q_1 : max at B_1, min at any point of Q_1 which lies on the circle $x^2 + y^2 = (3.4)^2$; Q_2 : max at $(0, 0)$, min on same circle as for Q_1.

■**A–5.4.** The point of D closest to $(0, 0)$ is at distance 1.4 from $(0, 0)$, and the farthest one is at distance 6.5. Therefore the values of $G(x, y)$ on D range from $G((1.4)^2 + 5)$ to $G((6.5)^2 + 5)$. In the interval $(1.4)^2 + 5 \leq t \leq (6.5)^2 + 5$ the max of $G(t)$ is 4 and the min is 1.

■**A–5.5.** a) Critical point $(0, 0)$; second derivative test is inconclusive. b) The level curve $f = 6$ consists of the three lines $x = 0$, $x - y = 0$ and $x + y = 0$. They subdivide the plane into six alternately green and red sectors which we describe with the θ of polar coordinates: $-45°$ to $45°$: green, $45°$ to $90°$: red, etc.   c) No. Any circle centered at $(0, 0)$, no matter how small the radius, contains green and red points. That is, $f(0, 0) = 6$ is not a local extremum.

■**A–5.6.** $(0, 0)$ is the only critical point, and the second derivative test is inconclusive. Draw a circle C around $(0, 0)$. At all points in the interior of C other than $(0, 0)$ the function takes on negative values (i.e., values less than $f(0, 0)$). This is true no matter how small the circle is. Therefore the origin is a local maximum.

■**A–5.7.** The level surfaces $f = k$ are planes which are all perpendicular to $\vec{n} = (3, -1, 2)$. If k increases from -500 to 500, the plane $f = k$ moves in parallel fashion in the direction of $\vec{n} = \text{grad} f$. It touches the sphere S for the first time at the point P on S for which $\overrightarrow{OP}$ is parallel to $-\vec{n}$, i.e., at the point $P(-6/\sqrt{14}, 2/\sqrt{14}, -4/\sqrt{14})$, and it touches S for the last time at the "opposite" point Q (see schematic Fig. A–5.7). Thus f has the global minimum at P and the global maximum at $Q(6/\sqrt{14}, -2/\sqrt{14}, 4/\sqrt{14})$.

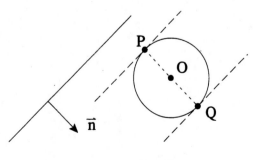

Fig. A–5.7

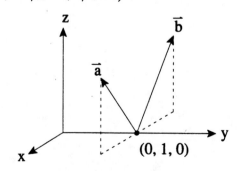

Fig. A–5.8

■ **A–5.8.** Think of drawing a solid ball B with center at $P(0, 1, 0)$. If you move away from P into the ball B in the direction of $\vec{a} = (1, 0, 1)$, then f takes on positive values as soon as you have left P (see Fig. A–5.8). Similarly, if you move out in the direction of $\vec{b} = (-1, 0, 1)$, the values will be negative. The ball B contains points where the values of f are higher than $f(P)$, and it contains points where the values are lower. This is true no matter how small I draw B. That is, the function f has neither a maximum nor a minimum at P.

■ **A–5.9.** (i) Write i for increase and d for decrease. Then A: $d - i - d$, B: $i - d - i$; C: $d - i - d - i$; D: i; E: i.
(ii) See Fig. A–5.9: $A_1, A_2, \ldots$ are the critical points of f along A, etc.
(iii) Local extrema: A_1 min, A_2 max; C_1 min, C_2 max, C_3 min; global extrema: B_1 max, B_2 min; no critical point of f on D; at E_1 no extremum.

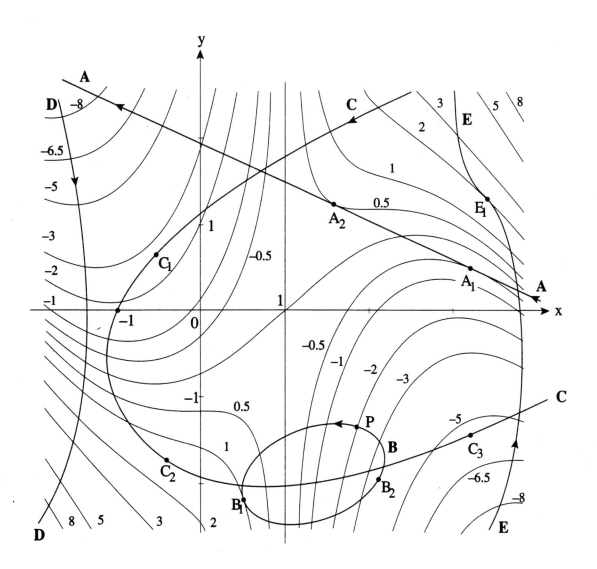

Fig. A–5.9

■**A–5.10.** a) min at $(4,0)$; b) min at $(0,1)$; c) min at $(-0.4, 0.2)$; d) min at $(0.3, -0.1)$; e) min at $\pm(\sqrt{3}, \sqrt{3})$; f) min at $(0,0)$. All extrema are local.

■**A–5.11.** a) none; b) critical point for $t=2$, neither min nor max.

■**A–5.12.** Critical points in terms of θ of polar coordinates: $\theta = 0$ max, $\frac{\pi}{4}$ local min, $\frac{\pi}{2}$ max, π min, $\frac{5\pi}{4}$ local max, $\frac{3\pi}{2}$ min.

■**A–5.13.** The level curves of f are lines with slope -1. The critical points of f on K are the points where the tangent has slope -1: $(2, 0.5)$ min on C, $(2, 3.4)$ neither min nor max, $(3.3, 5)$ max on C.

■**A–5.14.** see Fig. A–5.14

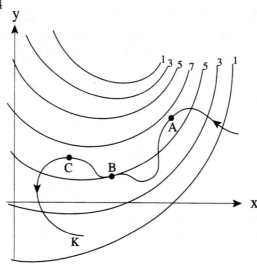

Fig. A–5.14

■**A–5.15.** a) (i) objective function $f(x,y,z) = y$, constraints $x^2 + xy + y^2 - 1 = 0$ and $x^2 + y^2 + z^2 - 16 = 0$; (ii) $F(x,y,z,\lambda,\mu) = y + \lambda(x^2 + xy + y^2 - 1) + \mu(x^2 + y^2 + z^2 - 16)$; b) (i) objective function $f(x,y) = xy$, constraint $y - (x-4)^2 = 0$; (ii) $F(x,y,\lambda) = xy + \lambda(y - (x-4)^2)$; c) (i) objective function $f(A,B,C) = (10A + 12C)/\sqrt{19}$, constraint $4A^2 + 4B^2 + 4C^2 - 1 = 0$; (ii) $F(A,B,C,\lambda) = (10A + 12C)/\sqrt{19} + \lambda(4A^2 + 4B^2 + 4C^2 - 1)$; d) (i) objective function $f(x,y,z) = (x-a)^2 + (y-b)^2 + (z-c)^2$, constraint $z - f(x,y) = 0$; (ii) $F(x,y,z,\lambda) = (x-a)^2 + (y-b)^2 + (z-c)^2 + \lambda(z - f(x,y))$.

6 | Polar, Cylindrical and Spherical Coordinates

■ **A–6.1.** a) K_1 is bounded by the x-axis and by parts of the spiral $r = \theta$ and of the circle $x^2 + y^2 = 1$; see Fig. A–6.1 a).
b) The curve bounding D on the right has the equation $r = (\cos\theta)^{-1}$ in the $r\theta$-plane; see Fig. A–6.1 b).

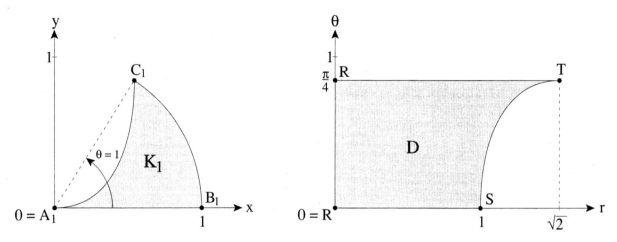

Fig. A–6.1 a) and b)

■ **A–6.2.** a) 0.0179 (obtained by taking $r = 6.31$ in $r\,\Delta r\,\Delta\theta$); b) 0.0180 ("exact" value).

■ **A–6.3.** a) R_1: $1.55 \leq r \leq 2.1$, $\pi \leq \theta \leq 7\pi/4$;

b) R_2: $0 \leq r \leq b/\cos\theta$, $0 \leq \theta \leq \pi/4$;

c) R_3: $0 \leq r \leq 2a\cos\theta$, $-\pi/2 \leq \theta \leq \pi/2$.

■ **A–6.4.**

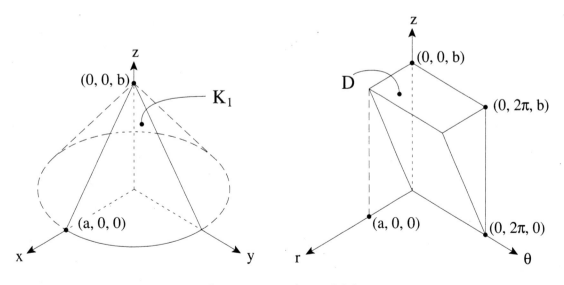

Fig. A–6.4 a) and b)

■ **A–6.5.** D_1: $b \leq r \leq a$, $\pi/2 \leq \theta \leq 2\pi$, $0 \leq z \leq h$;
D_2: $0 \leq r \leq (a/h)(h-z)$, $0 \leq \theta \leq 2\pi$, $0 \leq z \leq h$;
D_3: $0 \leq r \leq \sqrt{a^2 - z^2}$, $0 \leq \theta \leq 2\pi$, $-a \leq z \leq 0$.

■ **A–6.6.** a) Solid half-sphere $x^2 + y^2 + z^2 = a^2$, $z \leq 0$;
b) If we set $\theta = 0$ in the formulas for spherical coordinates we obtain a map from the $\rho\phi$-plane to the zx-plane which defines polar coordinates in the zx-plane: $z = \rho \cos \phi$, $x = \rho \sin \phi$. We use this observation to consider the triangle OA_1B_1 in which the solid cone D_1 intersects the zx-plane (see Fig. A–6.6 b) (ii) below center; the z-axis is drawn horizontally because it plays the role of the "horizontal" variable in the polar coordinates (ρ, ϕ)). In the mapping defined by these polar coordinates, the triangle OA_1B_1 is the image of the figure $OABC$ in the $\rho\phi$-plane (the equation of the curved part AB is $\rho = b/\cos \phi$ – see Fig. A–6.6 b) (i) below left; cf. also problem **6.1 b)**).

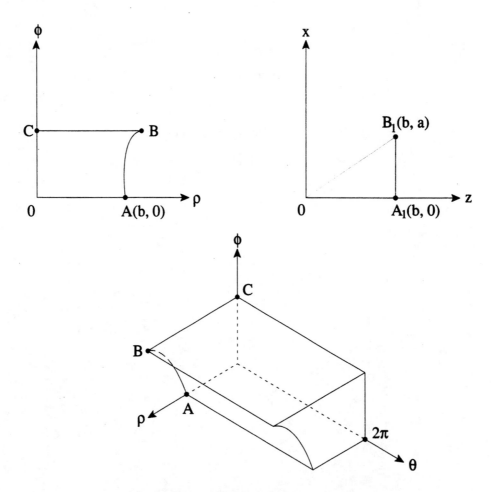

Fig. A–6.6 b) (i), b) (ii), and b) (iii)

As the triangle OA_1B_1 rotates around the z-axis to generate the cone D_1, the corresponding figure $OABC$ in $\rho\theta\phi$-space moves (in parallel fashion) along the θ-axis from 0 to 2π. Therefore the solid D must have the shape shown in Fig. A–6.6 b) (iii) above center.

■ **A–6.7.** D_1: $0 \leq \rho \leq a$, $\pi/2 \leq \phi \leq \pi$, $0 \leq \theta \leq 2\pi$;
D_2: $\rho_1 \cos\phi = b$ (see Fig. A–6.7 (ii)), and therefore $b/\cos\phi \leq \rho \leq a$, $0 \leq \phi \leq \arccos(b/a)$, $0 \leq \theta \leq 2\pi$;

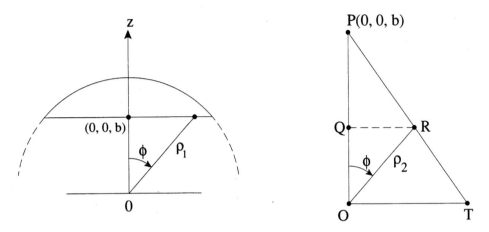

Fig. A–6.7 (ii) and (iii)

D_3: A halfplane $\theta = const$ intersects the cone in a triangle POT (see Fig. 6.7 (iii)). The triangles PQR and POT are similar. Therefore $\overline{PO}/\overline{OT} = b/a = \overline{PQ}/\overline{QR}$, and this equation yields a linear equation for ρ_2. $0 \leq \rho \leq ab/(b\sin\phi + a\cos\phi)$, $0 \leq \phi \leq \pi/2$, $0 \leq \theta \leq 2\pi$.

7 | Double Integrals

■ **A–7.1.** The subdivision is indicated by |, and the points chosen by •, with a) above and b) below the t-axis:

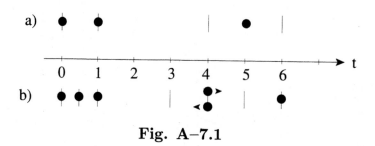

Fig. A–7.1

Sum of a): 2.9, sum of b): 1.1; c) $m = (-1.5) \cdot 6$, $M = 2.6 \cdot 6$.

■ **A–7.2.** a) See Fig A–7.2 a) below for first subdivision and points chosen; Riemann sum: $e^{2 \cdot 3} \cdot (3.375) + e^{(3.5) \cdot 3} \cdot (1.125)$;

b) See Fig. A–7.2 b) below for second subdivision and points chosen; Riemann sum: $e^{3 \cdot 2} \cdot (0.5) \cdot (3.375) + e^{(3.5) \cdot 3} \cdot (0.5) \cdot (1.125) + e^{(3.3) \cdot (3.5)} \cdot (0.5) \cdot (1.125) + e^{2 \cdot 3} \cdot (0.5) \cdot (3.375)$.

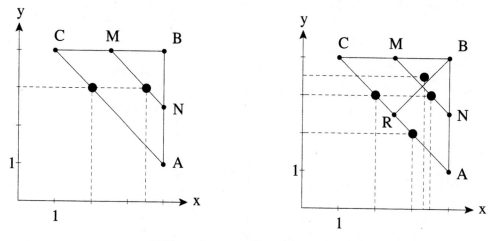

Fig. A–7.2 a) and b)

c) max on T: $e^{4.4}$, min: $e^{4.1}$, area: 4.5; $m = (4.5) \cdot e^4$, $M = (4.5) \cdot e^{16}$.

■ **A–7.3.** a) (i) $3\sqrt{2}$, (ii) $3\sqrt{2}$, $(1.5)\sqrt{2}$; (iii) counterclockwise, starting at A: $(1.5)\sqrt{2}$, 1.5, 1.5, $(1.5)\sqrt{2}$

b) See Fig. A–7.3 b) for S_1, S_2, S_3. (i) S_1: Subdivide T by drawing height of T; S_2: Subdivide each of the two triangles of S_1 by drawing the height in each of them; S_n: subdivide the 2^{n-1} triangles of S_{n-1} by drawing the height in each of them;

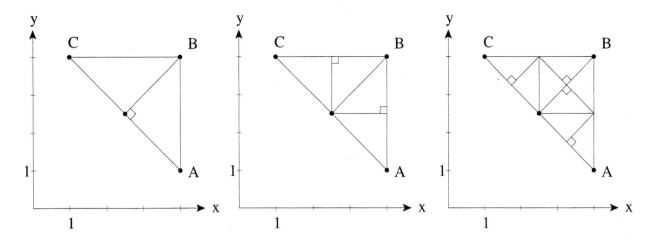

Fig. A-7.3b) S_1, S_2 and S_3

(ii) S_1: 3; S_2: $3 \cdot \frac{\sqrt{2}}{2}$; S_n: $3 \cdot (\frac{\sqrt{2}}{2})^{n-1}$

c) See Fig. A-7.3 c) for $\overline{S}_1$, $\overline{S}_2$, $\overline{S}_3$. (i) $\overline{S}_1$: subdivide T by drawing a parallel to AC through midpoint of CB; $\overline{S}_2$: subdivide each of the two pieces by drawing a parallel halfway through each; $\overline{S}_n$: subdivide each of the 2^{n-1} pieces of $\overline{S}_{n-1}$ by drawing a parallel halfway through each of them.

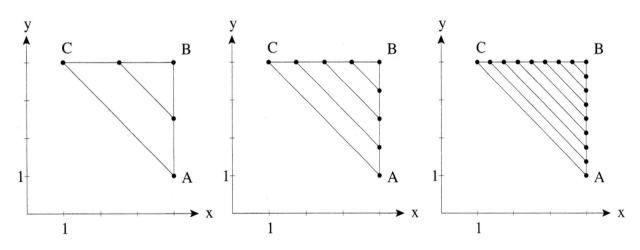

Fig. A-7.3 c) $\overline{S}_1$, $\overline{S}_2$ and $\overline{S}_3$

(iii) $\overline{S}_1$, $\overline{S}_2$, $\overline{S}_3$: always $3\sqrt{2}$.

■ **A-7.4.** Riemann sum: 6, integral: 6.

■ **A-7.5.** Riemann sum: 4.5, integral: 4.5.

■ **A-7.6.** The area is 5.5, and the global extrema of p in D are 0.4 and 3.5. Therefore $m = 2.2$ and $M = 19.2$.

■ **A–7.7.** a) t^* is approximately 2.1:

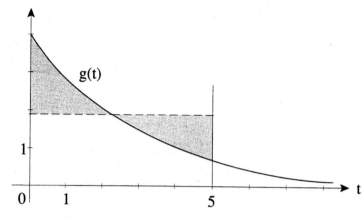

Fig. A–7.7

■ **A–7.8.** a) Formula language: $\iint_D f\, dA = f(x^*, y^*)\, A(D)$ where (x^*, y^*) is an appropriately chosen point in D and $A(D)$ is the area of D. – Geometry language: The volume under the surface $z = f(x, y)$ above D is the same as the volume of a straight prism whose base is D and whose height is $f(x^*, y^*)$ for an appropriately chosen point (x^*, y^*) in D.
b) The integral gives the volume of the triangular toy block shown in Fig. A–7.8 b). The block has the same volume as a rectangular box with base D and height $\frac{3}{2}$. For (x^*, y^*) any point on the straight line between $U(0, \frac{3}{2}, 0)$ and $V(1, \frac{3}{2}, 0)$ will work.

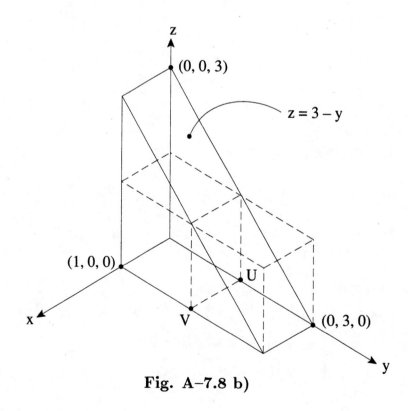

Fig. A–7.8 b)

■A–7.9. We combine the answers to a) and b).

R_1: $A(x,0)$, $B(x, 2-2x)$, $0 \le x \le 1$; $S(0,y)$, $T(1-\frac{y}{2}, y)$, $0 \le y \le 2$.

R_2: $A(x,-1)$, $B(x,1)$, $-2 \le x \le 1$; $S(-2,y)$, $T(1,y)$, $-1 \le y \le 1$.

R_3: $A(x,0)$, $B(x, \frac{x}{4})$, $0 \le x \le 4$; $S(4y, y)$, $T(4,y)$, $0 \le y \le 1$.

R_4: $A(x,0)$, $B(x, \sqrt{4-x^2})$, $-2 \le x \le 2$; $S(-\sqrt{4-y^2}, y)$, $T(\sqrt{4-y^2}, y)$, $0 \le y \le 2$.

R_5: (i) if $-2 \le x \le 0$: $A(x,0)$, $B(x, 2+x)$; (ii) if $0 \le x \le 2$: $A(x,0)$, $B(x, 2-x)$. Always $S(-2+y, y)$, $T(2-y, y)$, $0 \le y \le 2$.

R_6: $A(x,-2)$, $B(x,x)$, $-2 \le x \le 1$; $S(y,y)$, $T(1,y)$, $-2 \le y \le 1$.

R_7: $A(x,-\sqrt{x})$, $B(x, \sqrt{x})$, $0 \le x \le 2$; $S(y^2, y)$, $T(2,y)$, $-\sqrt{2} \le y \le \sqrt{2}$.

c) R_5: $\frac{8}{3}$; R_6: 0.

■A–7.10. a) First with respect to x; b) $(28.5)\,k$.

■A–7.11. a) $\int_a^b (\sin x + 3) dx$; b) $\int_a^b dx \int_0^{\sin x + 3} dy$.

■A–7.12. We write f for $f(x,y)$ and F for $f(r\cos\theta, r\sin\theta)$.

(1) $\int_{-2}^0 dx \int_x^{-x} f\,dy$; (2) $\int_{3\pi/4}^{5\pi/4} d\theta \int_0^2 F r\,dr$; (3) $\int_0^2 dx \int_{-2}^2 f\,dy - \int_0^1 dx \int_{-1}^1 f\,dy$;

(4) $\int_{-\pi/2}^{\pi/2} d\theta \int_1^2 F r\,dr$; (5) $\int_0^4 dx \int_0^2 f\,dy$; (6) $\int_0^{\pi/2} d\theta \int_0^{4\cos\theta} F r\,dr$.

8 | Triple Integrals

■ **A–8.1.** a) We subdivide the cylinder vertically into three equal cylinders of height 2 each. As points in the first and second we pick $(0, 4, 2)$, and $(0, 4, 4)$ in the third. The value of the Riemann sum is $8\pi(2 \cdot 20^{-3} + 32^{-3})$.
b) $m = 24\pi \cdot 72^{-3}$, $M = 24\pi \cdot 4^{-3}$.

■ **A–8.2.** $m = \frac{32}{3}\pi(4 - 2\sqrt{14})$, $M = \frac{32}{3}\pi(4 + 2\sqrt{14})$.

■ **A–8.3.** a) $0 \le z \le \frac{1}{4}(x-4)^2$, $0 \le y \le 6$, $0 \le x \le 4$; $W(x)$ is shown in Fig. A–8.3 a) below.

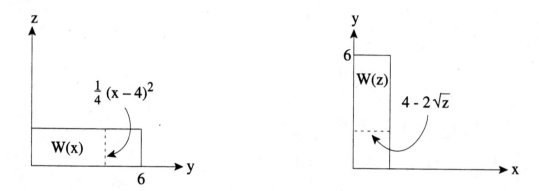

Fig. A–8.3 a) and c)

b) $\int_0^4 dx \int_0^6 dy \int_0^{\frac{1}{4}(x-4)^2} f(x,y,z)\,dz$

c) "Pick a z, say $z = c$. Intersect D with the plane $z = c$ and call the intersection W. The shape of W depends on the x chosen, and for that reason we write $W(z)$ instead of W; make a separate drawing of $W(z)$ in an xy-system (see Fig A–8.3 c));

describe $W(z)$ as a region in the xy-plane as per **7.9.**: Pick a $y, \dots$. Note that the x-values at which you enter and leave $W(z)$ may depend on the z and y chosen so far, and the y-values for the y-range of $W(z)$ may depend on z;

finally determine the range of z for D: Move the plane $z = c$ in the positive z-direction and call z_1 and z_2 the z-values at which you intersect D for the first and last time. This produces a description of D in the form

$$x_1(y,z) \le x \le x_2(y,z), \quad y_1(z) \le y \le y_2(z), \quad z_1 \le z \le z_2."$$

The integral is set up as $\int_0^4 dz \int_0^6 dy \int_0^{2\sqrt{z}+4} f(x,y,z)\,dx$, and $W(z)$ is shown in Fig. A–8.3 c) above.

d) $\int_0^6 dy \int_0^4 dx \int_0^{\frac{1}{4}(x-4)^2} f(x,y,z)\,dz$

■ **A–8.4.** D_1: Write p for $\frac{3(5-x)}{5}$. Then the integral is $\int_0^5 dx \int_0^{2p} dz \int_0^{p-\frac{z}{2}} f(x,y,z)dy$.

D_2: $\int_0^5 dy \int_{-2}^2 dx \int_0^{\sqrt{4-x^2}} f(x,y,z)dz$

D_3: $\int_0^6 dz \int_0^2 dy \int_{\frac{3}{2}y-3}^{4-2y} f(x,y,z)dx$

D_4: Write q for $\sqrt{\frac{y^4}{64} - x^2}$. Then the integral is $\int_0^4 dy \int_{-\frac{1}{8}y^2}^{\frac{1}{8}y^2} dx \int_{-q}^q f(x,y,z)dz$.

■ **A–8.5.** a) $\int_0^h dz \int_0^{2\pi} d\theta \int_0^{\frac{a}{h}(h-z)} (r\cos\theta \cdot r\sin\theta + z) r\, dr$

b) $\int_0^{2\pi} d\theta \int_0^{\frac{\pi}{2}} d\phi \int_0^{ah(h\sin\phi + a\cos\phi)^{-1}} (\rho\sin\phi\cos\theta \cdot \rho\sin\phi\sin\theta + \rho\cos\phi)\rho^2 \sin\phi\, d\rho$

■ **A–8.6.** a) In the formula for the approximate volume we take $\rho = 2.87$ and $\phi = \frac{\pi}{2}$. Then $\Delta V \approx 0.002\,279\,374\,218\,16$.

b) $\int_{1.48}^{\frac{\pi}{2}} d\theta \int_{1.52}^{\frac{\pi}{2}} d\phi \int_{2.87}^{2.93} \rho^2 \sin\phi\, d\phi = 0.002\,326\,357\,974\,84$.

9 | Vector Fields

■ **A–9.1.** a) Take an xyz-system such that $\vec{F_0} = 3\vec{k}$. Then $(F_1, F_2, F_3) = (0,0,3)$. b) Take an xyz-system such that $\vec{F_0}$ is parallel to $\vec{i} + \vec{j}$. Then $(F_1, F_2, F_3) = \frac{3}{\sqrt{2}}(1,1,0)$. c) $f(x,y,z) = 3z + k$ where k is any arbitrary constant. That is, there are infinitely many such functions f.

■ **A–9.2.** a) Set up the coordinate system such that A has coordinates $(0,0,0)$ and B has coordinates $(6,0,0)$. Then $\vec{F} = (2x-6, 2y, 2z)$. b) See Fig. A–9.2; $\vec{F}(M) = \vec{0}$. c) "Q is a critical point of f" means "$\vec{F}(Q) = \vec{0}$." Thus M is a critical point of f.

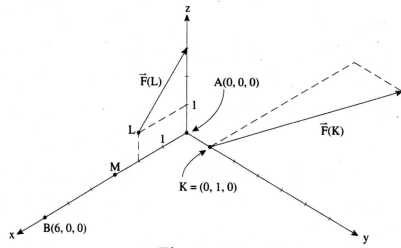

Fig. A–9.2

■ **A–9.3.** a) $x = 3\cos(\frac{\pi}{2}t)$, $y = 3\sin(\frac{\pi}{2}t)$. b) and c) see Fig. A–9.3. d) $\frac{\pi}{2}(-y, x)$.

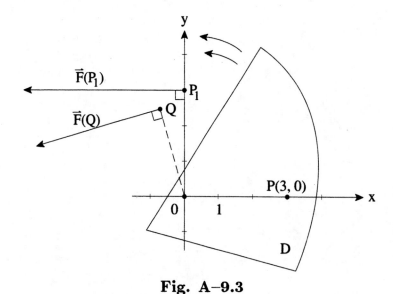

Fig. A–9.3

■ **A–9.4.** a) See Fig. A–9.4. b) $\vec{w}$ is not defined at the origin. $\vec{v} = \vec{0}$ at the origin, $\vec{u} = \vec{0}$ at the origin and on the circle $x^2 + y^2 = 1$. c) $0, \frac{\pi}{2}, -\frac{\pi}{2}$.

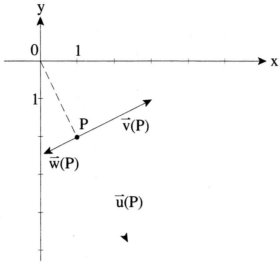

Fig. A–9.4

■ **A–9.5.** a) $\frac{g'(r)}{r}\vec{r}$. b) Each field vector is perpendicular to the sphere, and they all have the same length $g'(11) = 0.7$. c) As P moves, the field vector at P is always perpendicular to the sphere, and its length stays constant. d) $grad\, f$ is zero for $r = 4$. Take M and N on $x^2 + y^2 + z^2 = 16$, e.g., $(4,0,0)$ and $(0,4,0)$.

■ **A–9.6.** Flow line of $\vec{u}$: Ray $y = -2x$, $x \geq 0$; orientation away from the origin at points Q for which $|\overrightarrow{OQ}| > 1$, opposite if $|\overrightarrow{OQ}| < 1$. Flow line of $\vec{v}$: $x^2 + y^2 = 5$ counterclockwise. Of $\vec{w}$: $x^2 + y^2 = 5$ clockwise.

■ **A–9.7.** a) See Fig. A–9.7. b) Vertical lines, orientation up if $y < 0$ and down if $y > 0$. c) (i) $x = 4$, $y = 2e^{-t}$; (ii) see Fig. A–9.7; (iii) the point moves on the line $x = 4$ in the negative y-direction but stays all the time above the x-axis. For $t = -10^{189}$ it is way up, and for $t = 10^{189}$ it is extremely close to the x-axis but still above it.

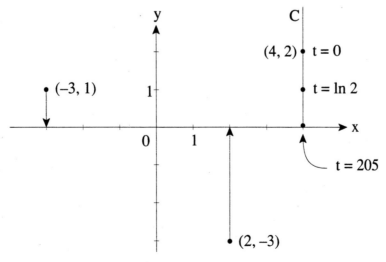

Fig. A–9.7

■ **A–9.8.** a) Right angle (because $grad\, f(x,y)$ is perpendicular to level curves of $f(x,y)$). b) Eight points are given. The level curves (broken) through the first four are hyperbolas $xy = \pm 1$, and through the last four they are the coordinate axes. The flowlines are perpendicular to the level curves, and the tangents to the flowlines through the eight points are shown as solid line segments. (Further computations show that the flowlines in question are the curves $x^2 - y^2 = k$ for $k = \pm 1, 0$, shown dotted.)

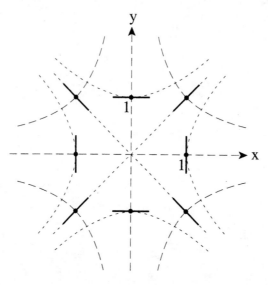

Fig. A–9.8

■ **A–9.9.** $\vec{b}$ must be perpendicular to the field vector at $(2, 4, 5)$, and this field vector has components $(0, 4, 1)$. For example, $\vec{b} = (1, 0, 0)$ will do.

■ **A–9.10.** A: (i) right hand term is nonsense because a and b are constants; B: (i) no, (ii) yes; C: (i) no, (ii) no; D: (i) both terms are nonsense because F_1 and F_2 are functions of *two* variables, and u', v' are functions of *one* variable; E: (i) no, (ii) yes.

■ **A–9.11.** See Fig. A–9.11.

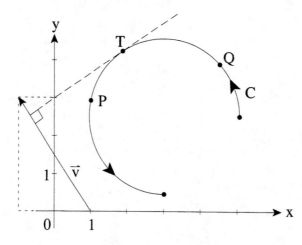

Fig. A–9.11

■ **A–9.12.** a) See Fig. A–9.12. b) First negative, then positive.

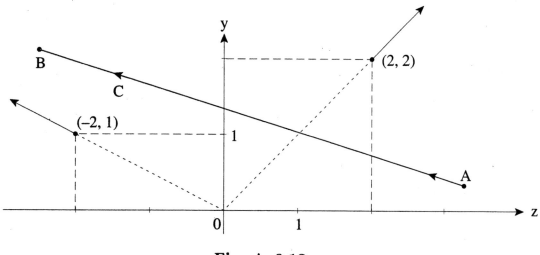

Fig. A–9.12

■ **A–9.13.** $(8t - 6t^3 + t^5)(9t^4 + 1 + 16t^2)^{-1/2}$

■ **A–9.14.** a) See Fig. A–9.14. b) (i) $|\vec{F}(Q)|$, (ii) 0. c) See Fig. A–9.14. Method 1 (trial and error): Draw any straight line through Q and change it around until the projection of the field vector on the line equals 2.5 units. Then draw a curve C which has the line as tangent at Q. Method 2: Use the fact from High School geometry that an angle inscribed in a semicircle is a right angle.

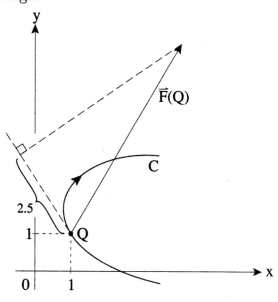

Fig. A–9.14

■ **A–9.15.** a) Draw the basis vectors $\vec{e_r}$ and $\vec{e_\theta}$ at P, and decompose $\vec{F}$ along these two vectors: $\vec{F} = F_r \vec{e_r} + F_\theta \vec{e_\theta}$; $F_r = -1.5$, $F_\theta = 2.6$. See Fig. A–9.15. b) and c): $F_r = -3\sin\theta$, $F_\theta = 3\cos\theta$.

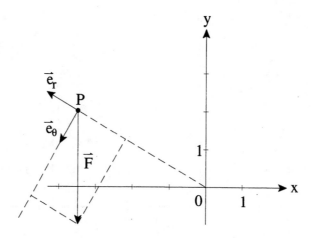

Fig. A–9.15

■ **A–9.16.** See Fig. A–9.16.

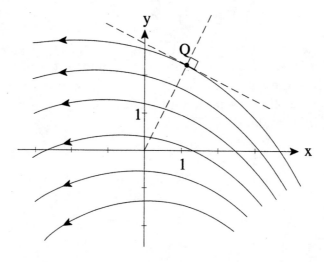

Fig. A–9.16

■ **A–9.17.** (i) $3x \cdot \vec{i} + 3y \cdot \vec{j} + 0z \cdot \vec{k}$; (ii) $3r \cdot \vec{e_r} + 0 \cdot \vec{e_\theta} + 0 \cdot \vec{e_z}$; (iii) $3\rho \sin\phi \sin\phi \cdot \vec{e_\rho} + 0 \cdot \vec{e_\theta} + 3\rho \sin\phi \cos\phi \cdot \vec{e_\phi}$.

10 | Line Integrals

■ **A–10.1.** a) See Fig. A–10.1, Riemann sum is -14.4; b) maximum value of f : 0.4, minimum: -5.6; $k = -52.6$, $K = 3.8$.

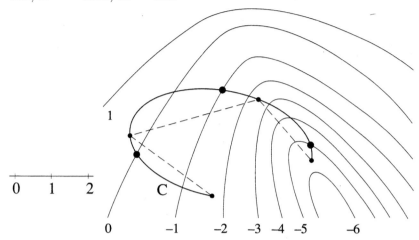

Fig. A–10.1

■ **A–10.2.** Maximum value of f: 2.0, minimum: 1.2; $k = 7.3$, $K = 12.6$

■ **A–10.3.** a) $15\sqrt{3}$; b) $-15\sqrt{3}$; c) the ordinary integral $\int_1^8 g(t)\,dt$ of elementary calculus.

■ **A–10.4.** a) $\int_0^\pi t^2 \sin t \cos t \sqrt{1+t^2}\,dt$; b) $\frac{\sqrt{3}}{4}\int_0^2 t^2(t^3+1)\sqrt{1+9t^4}\,dt$;
c) $\int_0^{\pi/2} \sin^3 t \cos^2 t \sqrt{1+\sin^2 t}\,dt$

■ **A–10.5.** a) See Fig. A–10.5, Riemann sum is 1.2; b) maximum value of F_{tan}: 5.1, minimum -4.3, $k = -40.5$, $K = 48.1$.

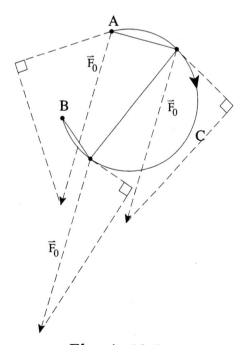

Fig. A–10.5

- **A–10.6.** In this case $F_{tan} = -|\vec{F}|$, maximum -0.5 and minimum -1, $K = -\frac{7\pi}{12}$, $k = -\frac{7\pi}{24}$.

- **A–10.7** $\int_{3\pi/2}^{0} (t\sin t + t\cos t + \sin t \cos t)dt$

- **A–10.8.** 32.25

- **A–10.9.** The ordinary integral $\int_1^5 f(t)dt$ of elementary calculus.

- **A–10.10.** a) $(9 \cdot e^9, 7)$ at all three points; b) -60.

- **A–10.11.** a) $\frac{f'(\sqrt{94})}{\sqrt{94}}(3, 9, -2)$; b) zero because C lies on the sphere $x^2 + y^2 + z^2 = 324$ and the field vector at P is parallel (or opposite) to $\vec{OP}$, i.e., perpendicular to the sphere.

- **A–10.12.** (3), (5), (6), (8), (9), (10).

- **A–10.13.** 4

- **A–10.14.** a) Take a straight line connecting any two points A, B on one of the level surfaces of f, e.g. $A(0,0,0)$ and $B(0,1,2)$ on the plane $3x - 4y + 2z = 0$; b) line connecting two points A, B on two different level surfaces, e.g. $A(0,0,0)$ and $B(1,1,1)$; c) impossible because $\vec{F}$ is conservative; d) any closed curve, e.g. $x = \cos t$, $y = \sin t$, $z = 0$.

- **A–10.15.** a) Integrals over C_1: 0, $\frac{3\pi}{2}$, 0; over C_2: 0, 2π, -2π; b) $\vec{u}$: impossible to decide; $\vec{v}$ and $\vec{w}$: no.

- **A–10.16.** No, because the *curl* is different from zero.

- **A–10.17.** Yes, $\vec{F} = grad(\frac{1}{4}r^4)$.

- **A–10.18.** The *curl* is zero. None of the partial derivatives of the components is defined at the origin. The test is inconclusive. However the integral around C_2 of **10.15** is different from zero, and therefore $\vec{F}$ cannot be conservative.

- **A–10.19.** a) The *curl* is zero, but the test is inconclusive. b) In order to avoid confusing the endpoint (x, y) with a point (x, y) anywhere on C, call the endpoint (a, b) and change back to (x, y) after you have evaluated the integral. The value of the integral is $\sqrt{x^2 + y^2} - 1$, and the partial derivatives of this expression are exactly the components of the given vector field. c) Yes, $\sqrt{x^2 + y^2}$.

- **A–10.20.** (1):10.16, (2):10.17, (3a):10.19, (3b):10.18.

- **A–10.21.** a) $\vec{F} = (1, f'(x))$; b) (i) a linear function, e.g., $f(t) = 2 - t$, (ii) for example $f(t) = t^2$.

- **A–10.22.** a) 0; b) $-x$; c) there is confusion on two issues. First, your colleague thinks that the following statement is true: "If the integral around a closed curve is zero (assumption), then the *curl* is zero (conclusion)." This statement is wrong as the present example shows. Secondly, the following theorem is true: "If the integral around **all** closed curves is zero, then the field is conservative." However this true theorem cannot be applied in our situation because we have not shown that the integral is zero over *all* closed curves. We have seen that the integral is zero over **one** particular closed curve.

■ **A–10.23.** a) 0; b) Your colleague forgot the part of the *curl*-test which deals with the partial derivatives of the components of $\vec{F}$: You compute the *curl* and find that it is zero. However, *if there is just one point at which one of the partial derivatives of a component is not defined*, then the test is inconclusive. In such a situation, the field may or may not be conservative. The field we consider is not conservative.

11 | Flux and Circulation in the Plane, Green's Theorem

■ **A-11.1.** See Fig. A-11.1.

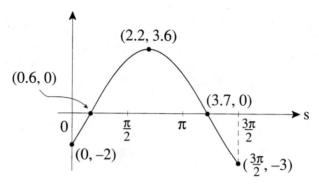

Fig. A-11.1

■ **A-11.2.** $\frac{-18}{\sqrt{17}}$

■ **A-11.3.** a) −11.8; b) $\overrightarrow{KL} = (0.5, -1.95)$.

■ **A-11.4.** (i) 13; (ii) −24.

■ **A-11.5.** a) Field vector at both L and M: $(0, 0.45)$, at K: $(0, 4.15)$; b) and c): flux across OP is $-\int_0^b f(x)dx$, across PQ: 0, across QO it will be positive. Computation of the latter: $\int_0^{b\sqrt{2}} (\sqrt{2}/2) f(t\sqrt{2}/2) dt = \int_0^b f(u) du$ which is the negative of the flux across OP. The total flux is zero.

■ **A-11.6.** The normal component of $\overrightarrow{F}$ along C equals −1 (note that C is oriented clockwise so that the normal points to the inside of C), and therefore the flux across C is $-2\pi b$. The normal component of $\overrightarrow{G}$ equals zero on C, and the flux is zero.

■ **A-11.7.** $\operatorname{curl} \overrightarrow{F} = f'(x)$, circulation: $bf(b) - \int_0^b f(y) dy$; $\operatorname{div} \overrightarrow{F} = 0$, and so the flux is zero.

■ **A-11.8.** For example, $a = b = m = 1$, $n = -1$, c and p arbitrary.

■ **A-11.9.** a) Zero for both because $\operatorname{div} \operatorname{grad} f = 0$; b) zero because $\overrightarrow{F}$ is conservative by definition.

■ **A-11.10.** a) $-k\Delta\theta$; b) the flux across the radial parts AB and CD is zero, across BC it is $k\Delta\theta$. That is, the flux across $ABCD$ is zero. c) Yes, because $\overrightarrow{F}$ is defined for all points in $ABCD$; $\operatorname{div} \overrightarrow{F} = 0$, and therefore the flux is zero. d) Green's theorem can be applied as in c); flux is zero. e) Green's theorem cannot be applied because $\overrightarrow{F}$ is not defined at the origin. We have to compute the flux and find $2\pi k$. f) Consider the following curve C (see Fig. A-11.10): C_3 once around from B to B, followed by BA, followed by the opposite of C_2, followed by AB.

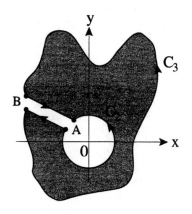

Fig. A–11.10 f)

This new curve C is the boundary of a domain which does not contain the origin. Therefore the flux across C vanishes. On the other hand,

$$\int_C = \int_{C_3} + \int_{BA} - \int_{C_2} + \int_{AB} = \int_{C_3} - \int_{C_2},$$

and we conclude that $\int_{C_3} = \int_{C_2}$. Since the latter flux was seen to be $2\pi k$ in e), the flux across C_3 equals $2\pi k$. g) Insert: "the origin," "zero," "zero," "$2\pi k$."

■ **A–11.11.** a) The flux is equal to the area of the parallelogram which is 13. b) The reason for the answer to a) is that $div\,\vec{F} = 1$. Insert "$\int_C -(y/2)dx + (x/2)dy$." c) The circulation of the vector field $(-\frac{y}{2}, \frac{x}{2})$ has to be computed for the curves I: $(a,0)$ to $(b,0)$ on the x-axis; II: $(b,0)$ vertically up to $(b, \sin b + 3)$; III: $(b, \sin b + 3)$ to $(a, \sin a + 3)$ backwards on $y = \sin x$; IV: $(a, \sin a + 3)$ vertically down to $(a, 0)$. The integrals over these curves are

$$\tfrac{1}{2}\int_a^b 0\,dt + \tfrac{1}{2}\int_0^{\sin b+3} b\,dt + \tfrac{1}{2}\int_b^a (-\sin t - 3 + t\cos t)\,dt + \tfrac{1}{2}\int_{\sin a+3}^0 a\,dt.$$

■ **A–11.12.** a) The circulation divided by area is $-2y_0 - b$ which tends to $-2y_0$, and the latter is $curl\,\vec{F}$ at (x_0, y_0). b) The flux divided by area is 1, and $div\,\vec{F} = 1$.

■ **A–11.13.** We abbreviate $\pi \cdot 10^{-6}$ by A. Circulation: $-3A/2$, $-4A$, 0; flux: 0, $20A$, $6A$.

12 | Surface Integrals, Flux Across a Surface

■ **A–12.1.** a) (i) Subdivision by strips $0 \le z \le 2$, $2 \le z \le 4$, $4 \le z \le 6$; (ii) points chosen: $(x, y, z) = (0, 3, 0), (0, 3, 2), (0, 3, 4)$; (iii) Riemann sum has value $8\pi(\frac{1}{9} + \frac{1}{13} + \frac{1}{25})$. b) minimum $\frac{1}{85}$ at $(0, 7, 6)$, maximum $\frac{1}{9}$ at $(0, 3, 0)$, $k = 24\pi\frac{1}{85}$, $K = 24\pi\frac{1}{9}$.

■ **A–12.2.** $k = 16\pi(4 - 2\sqrt{14})$, $K = 16\pi(4 + 2\sqrt{14})$.

■ **A–12.3.** $H(3) \cdot 36\pi = -124.41$

■ **A–12.4.** Plane of the parallelogram: $(x, y, z) = (1+3u+2v, 3u+v, 1+4u+v)$, $\iint_S f\,dA = \iint_D (16u + 5v + 1)\sqrt{35}\,du\,dv = \frac{23}{2}\sqrt{35}$ where D is the square $(0,0), (1,0), (0,1), (1,1)$ in the uv-plane.

■ **A–12.5.** $\iint_S f\,dA = \iint_D (x^2 + 4)2\,dx\,d\theta = 21\pi$ where D is the rectangle $(0,0), (3,0), (3, \frac{\pi}{2}), (0, \frac{\pi}{2})$ in the $x\theta$-plane.

■ **A–12.6.** We write M for $\frac{1}{\sqrt{35}}$. a) $-13M$; b) $-2M$, $-\frac{2}{7}M$; c) (i) $-63M$, (ii) $M(-36u - 14v + 29)$.

■ **A–12.7.** $\frac{16\sqrt{3}}{363}$

■ **A–12.8.** a) $(n_1, n_2, n_3) = (\cos u \sin v, \sin u \sin v, \cos v)$; b) $\frac{\sqrt{2}}{2}$, $(1, 4\cos^2 v, 0) \cdot (n_1, n_2, n_3)$.

■ **A–12.9.** (i) Subdivision by strips $0 \le z \le \frac{1}{2}$, $\frac{1}{2} \le z \le 1$, $1 \le z \le 2$; (ii) points chosen: $(2, 0, 0), (2, 0, \frac{1}{2}), (2, 0, 1)$; (iii) 2π.

■ **A–12.10.** The normal component has maximum 12 at $z = 3$, minimum 3 at $z = 0$, $k = 54\pi$, $K = 216\pi$.

■ **A–12.11.** a) See Fig. A–12.11; b) positive

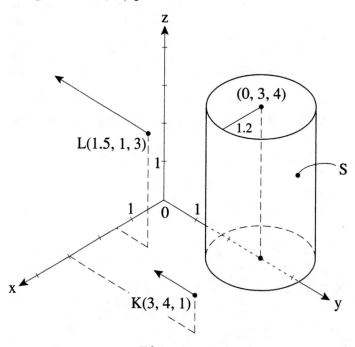

Fig. A–12.11

- **A–12.12.** The flux is $\frac{1}{2}\vec{F} \cdot (\vec{AC} \times \vec{AB}) = -57$.

- **A–12.13.** Looking down onto the xy-plane in the direction of $-\vec{k}$:

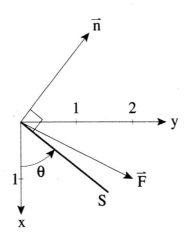

Fig. A–12.13

The flux is zero if $\tan\theta = 2$, i.e. $\theta = 1.107$ (radian measure); the flux is lowest for $\theta = 1.107 + \frac{\pi}{2}$.

- **A–12.14.** Flux across the face $z = 0 : 0$, $z = 4 : 0$, $y = 0 : -g(0) \cdot 8 = -36.8$, $y = 3 : g(3) \cdot 8 = -3.2$, $x = 0 : (-5) \cdot 12 = -60$, $x = 2 : 5 \cdot 12 = 60$, total -40.

- **A–12.15.** The normal component is b^{-2}, i.e., constant, on S. Flux: 4π.

- **A–12.16.** Plane of the triangle: $(x, y, z) = (2-2u-2v, 2u, 2v)$, flux $\iint_D (8-8v)dudv = \frac{8}{3}$ where D is the triangle $(0,0)$, $(1,0)$, $(0,1)$ in the uv-plane.

- **A–12.17.** S: $(x, y, z) = (2\cos t, 2\sin t, z)$, flux $\iint_D (8\cos t \sin t + 2\cos t)dtdz = 12$ where D is the rectangle $0 \le t \le \frac{\pi}{2}$, $0 \le z \le 2$ in the tz-plane.

- **A–12.18.** Flux $\iint_D (-r^3 + 8r)drd\theta = 24\pi$ where D is the rectangle $0 \le r \le 2$, $0 \le \theta \le 2\pi$ in the $r\theta$-plane.

- **A–12.19.** The normal component of $\vec{F}$ on S is $|grad\, f|$. Flux $\iint_T (4x^2 + 4y^2 + 1)dxdy = 29.5$ where T is the triangle in the xy-plane.

- **A–12.20.** The reasoning is incorrect. The theorem of Green deals with the flux of a vector field in the **plane** across a closed **curve** in the plane. Here we have to deal with a vector field in **space** across a **surface** (viz., the square $OABC$). Flux $\iint_D xydxdy = 4$ where D is the square $OABC$.

- **A–12.21.** The flux across the vertical wall is zero because $\vec{F}$ is perpendicular to the normal there. Flux across top: $5\iint_D x^2 dxdy = \frac{405}{4}\pi$ where D is the top disc.

13 | The Theorem of Gauss (Divergence Theorem)

■**A–13.1.** M_1: solid; surface or empty; the spherical shell $x^2 + y^2 + z^2 = 9$; $U(0,0,0)$ and $V(3,0,0)$.
M_2: surface; curve or empty; empty; $U(3,0,0)$, there are no points V.
M_3: surface; curve or empty; the circle $x^2 + y^2 = 9$ in the plane $z = 0$; $U(0,0,3)$ and $V(3,0,0)$.
M_4: solid; surface; surface M_3 plus disk $x^2 + y^2 \leq 9$, $z = 0$; $U(0,0,1)$ and $V(0,0,0)$.
M_5: surface; curve or empty; the circle $x^2 + y^2 = 4$, $z = 0$; $U(0,0,0)$ and $V(2,0,0)$.
M_6: curve; two points or empty; empty; $U(4,4,0)$, there are no points V.
M_7: curve; two points or empty; points $(5,0,0)$ and $(-5,0,0)$; $U(0,5,0)$ and $V(5,0,0)$.
M_8: surface; curve or empty; the curve consisting of the four line segments AB, BC, CD, DA where $A(2,1,0)$, $B(1,5,7)$, $C(4,4,9)$, $D(5,0,2)$; U corresponding to $(u,v) = (\frac{1}{2}, \frac{1}{2})$, i.e., $U(3, 2.5, 4.5)$, and $V(2,1,0)$ corresponding to $(u,v) = (0,0)$.

■**A–13.2.** ∂R consists of the two spherical shells $x^2 + y^2 + z^2 = 1$ and $x^2 + y^2 + z^2 = 4$. On the former the exterior normal points towards the origin, and on the latter away from the origin.

■**A–13.3.** a) ∂M_1: the circles $x^2 + y^2 = 4$, $z = 0$ and $x^2 + y^2 = 4$, $z = 5$;
∂M_2: the circle $x^2 + y^2 = 4$, $z = 0$;
∂M_3: S plus T plus bottom disk $x^2 + y^2 \leq 4$, $z = 0$;
∂M_4: the two points $(2,0,0)$ and $(2,0,5)$. b) S_1: the disk $x^2 + y^2 = 4$, $z = 2.5$;
S_2: T plus the upper half $2.5 \leq z \leq 5$ of S.

■**A–13.4.** a) The flux is zero because $\text{div}\,\vec{F} = 0$. b) Zero, as for a). c) The divergence of $\vec{G}$ has to vanish; take e.g. $\vec{G} = (y, z, x)$.

■**A–13.5.** K is the solid triangle $OABC$. ∂K consists of the four triangles OAB, OAC, OBC, and ABC. The flux across ∂K is zero because $\text{div}\,\vec{F} = 0$. Therefore the flux across the first three triangles must be $-\frac{8}{3}$ because the flux across the ABC was seen to be $\frac{8}{3}$.

■**A–13.6.** a) $4 \int_{1.8}^{4.2} 2 f'(y) \sqrt{1.44 - (x-3)^2}\, dy$; b) $4 \int_0^{2\pi} f(3 + 1.2 \sin t)(1.2) \sin t\, dt$; c) the graph $u = f(3 + 1.2 \sin t)(1.2) \sin t$ in Fig. A–13.6 shows that the integral in question is positive.

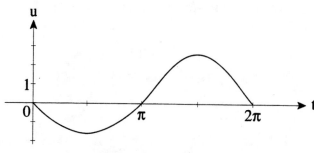

Fig. A–13.6

■ **A–13.7.** The flux equals $\iiint_K g'(y)dV = 8(g(3) - g(0)) = -40$ where K is the solid box.

■ **A–13.8.** The flux equals $\iiint_K 2z\,dV = 36\pi a^2$ where K is the solid cylinder.

■ **A–13.9.** a) 0 b) If one applied the divergence theorem one would have to integrate over a solid ball containing the origin. This cannot be done because $\vec{F}$, and therefore $div\,\vec{F}$, is not defined at the origin. c) According to **12.15** the flux across both S_1 and S_2 is 4π. If the divergence theorem were applicable - which it is not - the flux would have to be zero. d) The divergence theorem is applicable because $\vec{F}$ is defined for all points of K. The flux is zero because the divergence of $\vec{F}$ vanishes at all points of K.

■ **A–13.10.** a) $g(3)(1,1,1) = (2.7, 2.7, 2.7)$ b) (i) $3g(r^2)+2r^2g'(r^2)$; $(ii) 8.1+6\cdot(-0.7) = 3.9$. c) The normal component F_n is zero on N and $g(r^2)\cdot 2$ on B. $\iint_B 2g(x^2+4+z^2)dA = 2\pi\int_4^5 g(u)du = 2.5\cdot\pi$.

■ **A–13.11.** The flux across ∂K is zero because $div\,\vec{F} = 0$. Therefore the flux across B is -3.75.

■ **A–13.12.** K is the solid halfball. The flux across ∂K is $\frac{8}{3}\pi b^3$ by Gauss' theorem. At the same time the flux across ∂K equals the flux across S plus the flux across the bottom disk B. The latter is $-\frac{\pi}{2}b^4$. Therefore the flux across S is $\frac{8}{3}\pi b^3 + \frac{\pi}{2}b^4$.

■ **A–13.13.** a) There is no flux across the top and bottom; the flux across ∂K is $\int_{\partial D} -Q\,dx + P\,dy$. b) $\iint_D (P_x + Q_y)dx\,dy$ c) Green's theorem, in the divergence version, for the vector field in the plane with components $(P(x,y), Q(x,y))$.

■ **A–13.14.** a) $div\,curl\,(-y^2, x^2, z^2) = 0$; therefore the flux is zero. b) Insufficient information for the flux of $\vec{F}$ across ∂K_2 because $div\,\vec{F} = 2z$; flux of $curl\,\vec{F}$ is zero. c) Zero, no matter what the functions a, b, c are. d) "$curl\,\vec{G}$," "the boundary ∂K of a solid K," "zero."

14 | The Theorem of Stokes

- **A–14.1.** See A–13.1.
- **A–14.2.** $\operatorname{curl}\vec{F} = (-1,0,0)$, and the circulation equals $-\frac{3}{\sqrt{11}}\pi R^2$.
- **A–14.3.** $b^2\pi$
- **A–14.4.** a) $-\pi$; b) $\pi, -\pi, -\pi$; c) "$\operatorname{curl}\vec{G}$," "$S$," "$S^*$," "the same."
- **A–14.5.** a) zero; b) zero; c) "zero," "∂S," "$\operatorname{curl}\vec{G}$ across S is zero." d) $\operatorname{curl}\vec{H} = (6yz^2, -6xz^2, 0) \neq (0,0,0)$
- **A–14.6.** a) ∂T_1 is the circle C, with an orientation which looks clockwise in the drawing; b) $\int_C \vec{F}\cdot d\vec{r} = \iint_{T_1(b)} \operatorname{curl}\vec{F}\cdot d\vec{A}$; c) The limit of the left side is zero because a line integral over a curve consisting of one point is zero. The limit of the right side is the flux across the unmutilated torus T, for the following reasons. At the limit, T_1 is a torus with one point removed. The flux across T is the flux across T_1 plus the flux across this one point. The latter flux is zero because a surface integral over a surface consisting of one point is zero. d) "$\operatorname{curl}\vec{F}$," "$S$," "zero." e) Since S is a closed surface its boundary is empty. A line integral over an empty curve is zero.
- **A–14.7.** a) (i) $(0,0,0)$, (ii) $\vec{F}$ is defined everywhere except on the x-axis; b) 2π c) see Fig. A–14.7; A and B are the endpoints of the cut C.

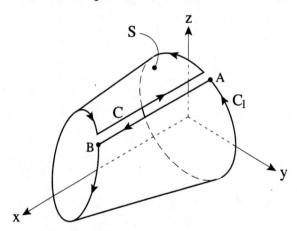

Fig. A–14.7

d) It is zero because $\vec{F}$ is defined for all points of S (S stays away from the x-axis!), and $\operatorname{curl}\vec{F}$ vanishes; e) $\int_{C_1} + \int_{AB} + \int_{-C_2} + \int_{BA} = 0$, and therefore $\int_{C_1} = \int_{C_2} = 2\pi$; f) $\vec{F}$ is defined for all points of the disk D, and $\operatorname{curl}\vec{F}$ vanishes. Therefore the circulation is zero; g) We cannot apply the reasoning of g) to C_2 because any surface S of which C_2 could be the boundary would have to have a point in common with the x-axis. On the latter, $\vec{F}$ is not defined, and therefore we cannot apply Stokes' theorem to such an S.

- **A–14.8.** Evaluation of the circulation along ∂S: $\int_{KL} = \int_{MN} = 0$, $\int_{LM} = cf(b)$, $\int_{NK} = -cf(a)$, total circulation $c(f(b) - f(a))$. Evaluation of flux: $\iint_S f'(x)dxdy = c\int_a^b f'(x)dx$. Thus, Stokes' theorem reduces to the fundamental theorem of elementary calculus.

15 | Gauss' and Stokes' Theorem, Miscellanea

- **A–15.1.** $\vec{F}$: flux 0, 0, 0, circulation 3π; $curl\,\vec{F}$: flux 0, 3π, 3π, circulation 0.
- **A–15.2.** (A)–(2), (B)–(5), (C)–none, (D)–none, (E)–(1), (F)–none.
- **A–15.3.** $-\frac{28}{3}\pi \cdot 10^{-9}$
- **A–15.4.** If $\vec{n}$ is a unit vector normal to α then $lim_{b\to 0}(\pi b^2)^{-2}\int_C \vec{F}\cdot d\vec{r} = curl\,\vec{F}\cdot \vec{n}$, that is, $\int_C \vec{F}\cdot d\vec{r} \approx (\pi b^2)\,curl\,\vec{F}\cdot\vec{n}$. Therefore we can take for α any plane whose normal is perpendicular to $(1,1,4)$ (or, equivalently, which contains the vector $(1,1,4)$), for example, $x-y=0$.

16 | Coordinate Transformations

■**A–16.1.** a) $A:(1,0)$, $B:(0,1)$, $O:(0,0)$, $P:(1.25,1.5)$, $Q:(-0.4,2.7)$; b) and c) see Fig. A–16.1.

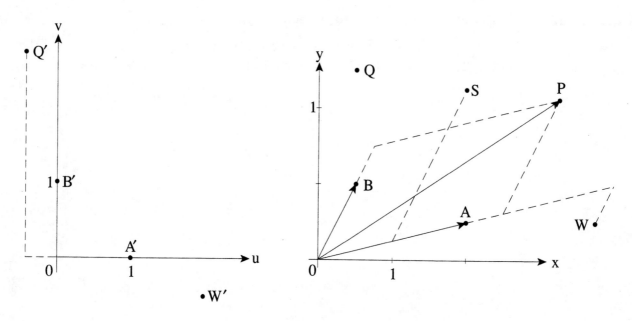

Fig. A–16.1 (i) and (ii)

d) $x = 2u + (0.5)v$, $y = (0.5)u + v$; $u = (4/7)x - (2/7)y$, $v = -(2/7)x + (8/7)y$.

■**A–16.2.** a) $x = 2u + v$, $y = -(1/2)u + (3/2)v$; b) $x = u + 2v$, $y = (3/2)u - (1/2)v$.

■**A–16.3.** $x = u\cos\phi - v\sin\phi$, $y = u\sin\phi + v\cos\phi$; see Fig. A–16.3.

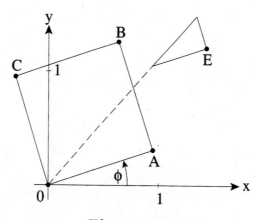

Fig. A–16.3

■ **A–16.4.** a) take $\phi = \pi/6$ in **16.3**; b) $-(1/2)x + (\sqrt{3}/2)y = ((\sqrt{3}/2)x + (1/2)y)^2$.

■ **A–16.5.** $x = u + 2v + w$, $y = 3u + 3v + w$, $z = 4u + 5w$

■ **A–16.6.** a) See Fig. A–16.6.

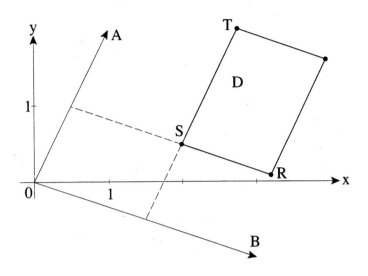

Fig. A–16.6

b) (i) $\Delta u \Delta v = 0.3$; (ii) $A(D) = |\overrightarrow{OA} \times \overrightarrow{OB}|\Delta u \Delta v = 2.1 = 7\Delta u \Delta v$; (iii) E' is the unit square in the uv-plane. (iv) We set $\Delta u = \Delta v = 1$ in (ii) and obtain $A(E) = 7$.

c) We expect $A(G) = 7A(G')$, arguing heuristically as follows. We cover G' with rectangles like D' considered in a), and this will generate a covering of G with parallelograms whose edges are parallel to $\overrightarrow{OA}$ and $\overrightarrow{OB}$. The sum of the area of the rectangles covering G' will be a first crude approximation $A_1(G')$ of $A(G')$ (it will be an approximation because some of the rectangles may contain points which are not in G'), and $A_1(G) = 7A_1(G')$ will be a first crude approximation of $A(G)$. If we use smaller rectangles (e.g., by subdividing each rectangle into 4 smaller ones), we get better approximations $A_2(G')$ and $A_2(G)$, and we still have $A_2(G) = 7A_2(G')$. Repeating this procedure of using smaller rectangles and passing to the limit we are likely to get $A(G) = 7A(G')$. This is only a sketch of a mathematical argument, of course.

d) There is a general pattern. If (a_1, a_2) and (b_1, b_2) are the xy-components of $\overrightarrow{OA}$ and $\overrightarrow{OB}$, then the area of the parallelogram spanned by these two vectors is $|a_1 b_2 - a_2 b_1|$. On the other hand, the formulas for the map **F** are

$$x = a_1 u + b_1 v, \quad y = a_2 u + b_2 v,$$

and the Jacobian determinant of this map is $a_1 b_2 - a_2 b_1$. This shows that the factor k in question is the absolute value of the Jacobian determinant.

■ **A–16.7.** a) 1; b) Formula reasoning: $A(E) = k \cdot A(E')$ where k is the absolute value of the Jacobian determinant which equals 1. – Geometric reasoning: The effect of the map is a rotation around the origin. Under such a rotation, the area of a figure does not change.

■ **A-16.8.** $x = u+v$, $y = u-v$; see Fig. A-16.8.

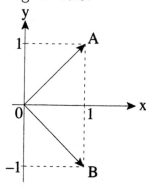

Fig. A-16.8

■ **A-16.9.** a) 19; b) -19;

c) Rewritten question c): Let G' be any solid of volume $V(G')$ in uvw-space, and let G the corresponding solid in xyz-space. What kind of relation would you expect between $V(G')$ and $V(G)$? – Answer to the rewritten question: We expect $V(G) = 19V(G')$, arguing heuristically as follows. We cover G' with rectangular boxes whose edges of length Δu, Δv and Δw are parallel to the uvw-axes. This will generate a covering of G with parallelepipeds whose edges are parallel to $\overrightarrow{OA}$, $\overrightarrow{OB}$ and $\overrightarrow{OC}$. The sum of the volume of the boxes covering G' will be a first crude approximation $V_1(G')$ of $V(G')$ (it will be an approximation because some of boxes may contain points which are not in G'), and $V_1(G) = 19V_1(G')$ will be a first crude approximation of $V(G)$. If we use smaller boxes (e.g., by subdividing each box into 8 smaller ones), we get better approximations $V_2(G')$ and $V_2(G)$, and we still have $V_2(G) = 19V_2(G')$. Repeating this procedure of using smaller boxes and passing to the limit we are likely to get $V(G) = 19V(G')$. This is only a sketch of a mathematical argument, of course.

Rewritten question d): You realize that the absolute value of the Jacobian determinant equals the constant factor k in the equation $V(G) = k \cdot V(G')$. Is this a coincidence, or is there a general pattern? – Answer to the rewritten question d): The volume of the parallelepiped is the absolute value of the mixed product of the vectors $\overrightarrow{OA}$, $\overrightarrow{OB}$ and $\overrightarrow{OC}$. The mixed product is computed as the determinant of the matrix

$$\begin{pmatrix} a_1 & a_2 & a_3 \\ b_1 & b_2 & b_3 \\ c_1 & c_2 & c_3 \end{pmatrix},$$

where $a_1, \ldots$ are the xyz-components of the vectors. This matrix is the transpose of the Jacobian matrix of the map, and the determinant does not change under transposition. Therefore $V(G) = k \cdot V(G')$ where k is the absolute value of the Jacobian determinant.

■ **A–16.10.** a) $f_1 = 1 + \frac{1}{2}(r-2) + (-\sqrt{3})(\theta - \frac{\pi}{3})$, $g_1 = \sqrt{3} + \frac{\sqrt{3}}{2}(r-2) + 1 \cdot (\theta - \frac{\pi}{3})$; b) see Fig. A–16.10.

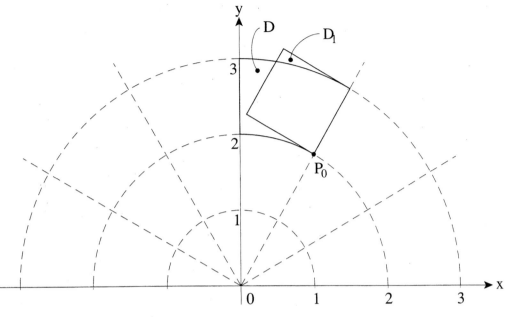

Fig. A–16.10

c) $A(D') = \frac{\pi}{6}$ (by definition of D'), $A(D_1) = 1.05 \cdot 1$ by measurement, and $k = \frac{1.05 \cdot \pi}{6} \approx 2$.
d) The Jacobian determinant equals 2. According to theory, the k of c) should be 2.

■ **A–16.11.** a) $\overrightarrow{P_0 K} = \Delta v(f_v, g_v)$, $\overrightarrow{P_0 M} = \Delta u(f_u, g_u)$ where $f_v, g_v \ldots$ are evaluated at (u_0, v_0). b) (i) $A(D_1) = \Delta u \Delta v \cdot k$ where $k = |f_v g_u - f_u g_v|$; (ii) k equals the absolute value of the Jacobian determinant $f_u g_v - f_v g_u$; (iii) "$\Delta u \Delta v$ times the absolute value of the Jacobian determinant of $\mathbf{F}$ at P_0'" because the area of the parallelogram D_1 is approximately the area of the curved figure D.

■ **A–16.12.** a) The Jacobian determinant equals e^{-2u}; note that it is independent of v. (i) $u = 0$, (ii) $u = 1.63$, (iii) $u = -2.40$. b) See Fig. A–16.12.

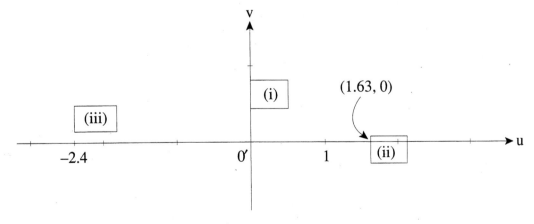

Fig. A–16.12

■ **A–16.13.** a) See Fig. A–16.13.

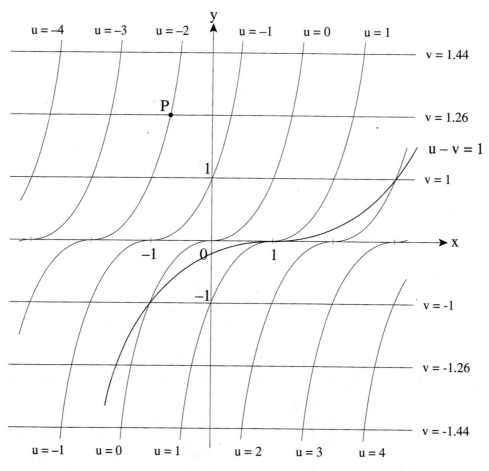

Fig. A–16.13

b) It is likely to be positive and greater than 1, for the following reasons. The equation of the line $v = 1.26$ looks like $x = u$, $v = 1.26$, and therefore $\vec{r_u} = (f_u, g_u)$ is likely to be of length 1 at P. Since an increase of v from 1.26 to 1.44 causes the arc length on $u = -2$ to increase by much more than 1, we expect $\vec{r_v}$ to be *much* longer than 1. Therefore the parallelogram spanned by $\vec{r_u}$ and $\vec{r_v}$ will have area greater than 1, which means that the absolute value of J will be greater than 1. In order to figure out the sign of J we observe that $\vec{r_v} = (f_v, g_v)$ points upwards (in the direction of increasing parameter values). Therefore J is positive (look at J as the third component of $\vec{r_u} \times \vec{r_v}$, with both vectors considered as vectors in space).

c) At any point on the x-axis, the tangents to the curves $u = constant$ and $v = constant$ are parallel. Therefore $\vec{r_u} \times \vec{r_v}$ vanishes on the x-axis. Since $|J| = |\vec{r_u} \times \vec{r_v}|$ we conclude that $J = 0$ on the x-axis.

d) (4)

■ **A–16.14.** a) (i) $u = 3$: surface generated by rotating the curve $x = f(3, w)$, $y = 0$, $z = g(3, w)$ around the z-axis; (ii) $v = \frac{\pi}{6}$: halfplane $\theta = \frac{\pi}{6}$ where θ is the angle of the usual cylindrical coordinates; (iii) $w = 2$: surface generated by rotating the curve $x = f(u, 2)$, $y = 0$, $z = g(u, 2)$ around the z-axis.

b) J_1: Jacobian determinant of the uw-coordinates in the xz-plane; J: Jacobian determinant of the uvw-coordinates. Then $J = f(u,w) \cdot J_1$.
c) Usual cylindrical coordinates ($u = r$, $v = \theta$, $w = z$); $J_1 = 1$, $J = r \cdot 1 = r$.
d) Usual spherical coordinates ($u = \rho$, $v = \theta$, $w = \phi$); $J_1 = -u$, $J = (-u) \cdot u \sin w = -u^2 \sin w$. The negative sign arises because in our setup we obtain $\frac{\partial(x,y,z)}{\partial(r,\theta,\phi)}$ and not the usual $\frac{\partial(x,y,z)}{\partial(r,\phi,\theta)}$.

■ **A–16.15.** 12.5

■ **A–16.16.** The statement is wrong. The integrand $f(u,v)$ on the right side should be replaced by $f(\cos\phi \cdot u - \sin\phi \cdot v, \sin\phi \cdot u + \cos\phi \cdot v)$.

■ **A–16.17.** a) $\int_0^2 dv \int_1^3 (2u + v) \cdot 3 \cdot du$; b) $\int_{-3}^0 dy \int_0^{3+y} (2x - y + 5) \cdot (1/3) \cdot dx$; see Fig. A–16.17.

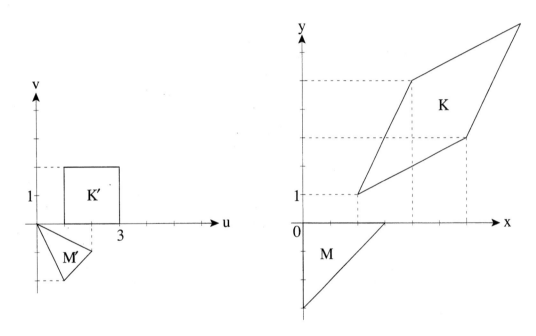

Fig. A–16.17

■ **A–16.18.** $\int_0^1 dv \int_0^1 \int_0^1 (4u + 5w) \cdot 19 \, du\, dw$

■ **A–16.19.** a) 28.21; b) $-\frac{1}{2}$; c) The solid T' in uvw-space corresponding to T is the tetrahedron with vertices $(u,v,w) = (0,0,0), (1,0,0), (0,1,0), (0,0,1)$; $\int_0^1 dw \int_0^{1-w} dv \int_0^{1-w-v} v \cdot 2 \cdot du$.

■ **A–16.20.** a) The curve $u = k$ is a parabola of the form $y = -x^2 + Ax + B$, and $v = k$ is a parabola of the form $y = x^2 + Cx + D$. b) $4ab$; c) $b((1+a)^2 - 1) + a((1+b)^2 - 1)$; d) the estimate is approximately 96.6% of the true value of the area.

■ **A–16.21.** a) $\int_a^b du \int_p^q |f_u g_v - f_v g_u| dv$

b) $\int_a^b du \int_p^q \sqrt{(g_u h_v - g_v h_u)^2 + (h_u f_v - h_v f_u)^2 + (f_u g_v - f_v g_u)^2} \, dv$

c) Projecting from xyz-space onto the xy-plane amounts to setting $h(u,v) \equiv 0$. If this is done in b) the integrand reduces to $\sqrt{(f_u g_v - f_v g_u)^2}$ which is the integrand of a).

■ **A–16.22.** a) See Fig. A–16.22.

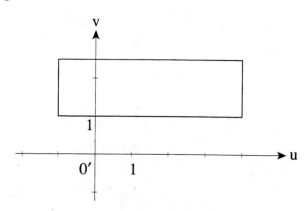

Fig. A–16.22

b) (i) $\frac{3}{2}\int_1^3 H(x)dx$; (ii) $\frac{3}{2}\int_{-1}^4 H(f(u))f'(u)du$ (note that $f'(u) \geq 0$ so that $|f'| = f'$). (iii) If you disregard the factor $\frac{3}{2}$ you obtain the substitution formula for ordinary integrals. Note that the limits of integration came out properly adjusted.

■ **A–16.23.** (i) $\int_b^{2b} dv \int_0^1 du \int_0^{1-u} (2u + 2v)dw = \frac{1}{3}b + \frac{3}{2}b^2$; (ii) $\frac{b}{2}$, the volume of K'.

■ **A–16.24.** a) $x = (A + r\cos\phi)\cos\theta$, $y = (A + r\cos\phi)\sin\theta$, $z = r\sin\phi$ where θ is the angle of rotation of the usual cylindrical coordinates. b) See Fig. A–16.24.

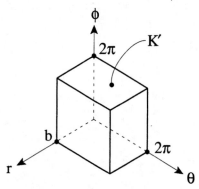

Fig. A–16.24

c) $J = r(A + r\cos\phi)$; d) We must have $A > r$ for all r because otherwise the torus would have no hole in the center. Consequently $J = r(A + r\cos\phi) > r(r + r\cos\phi) = r^2(1 + \cos\phi)$, and $1 + \cos\phi$ is never negative. Thus J is positive.

$\iiint_K z^2 dx\,dy\,dz = \int_0^{2\pi} d\phi \int_0^{2\pi} d\theta \int_0^b (r^2 \sin^2\phi)(Ar + r^2\cos\phi)dr$

Index

Answers:
    exact, 1
    "None," "Impossible" etc., 1
    to all problems, 135–184
Arc length, 6, 7, 8, 9, 10, 14
Area of a plane figure, 60, 89
Axonometry, isometric, 1–2

Boundary of surfaces and solids, 101

Chain rule, 27–28
    higher derivatives, 27
Change of variables in multiple integrals, 128–133
    linear transformations, 128–129
    non-linear transformations, 130–133
    and substitutions in ordinary integrals, 131–132
Circulation, ii, 77 (see also Ch. 11, 14, 15)
    Green's theorem for, 87
    per area, 90
Closed surface, 111
Constrained extrema, 44–47
Coordinate curves:
    of curvilinear coordinates, 122–128
    of a surface, 14, 16
Coordinate lines of a plane, 11, 13
Coordinate transformations, 117–133
    curvilinear coordinates, 122–128
        linear approximations, 122–125
    Jacobian of linear transformations, 121–122
    linear transformations, 117–122, 128–129
Curl (see Ch. 10, 11, 13, 14, 15)
    invariant definition, 90, 116
    test, 81–83

Curvilinear coordinates, 122–128
(see also Polar, Cylindrical, Spherical coordinates)
    linear approximation, 122–125
Cylindrical coordinates, 10, 15, 16, 51–52, 128
    interpretation as mapping, 51
    line integral over a curve in, 76
    triple integral in, 66

Directional derivative (see also Gradient), 29–36
    and average rate of change, 30
    in different directions at one point, 32
    used for estimate, 33
Divergence, ii (see Ch. 11, 12, 13, 15):
    invariant definition, ii, 90, 116
    theorem (theorem of Gauss), 102–107
        relation with Green's theorem, 106
Double integrals, 55–61
    estimates, 57
    mean value theorem, 58
    note on definition, 55
    Riemann sums, 56–57
    setting them up:
        in cartesian coordinates, 58–60
        in polar coordinates, 60–61

Flow lines of a vector field, 69–70
Flux, ii (see also Ch. 11–15)
    across a plane curve, 86–87
    across a surface, 95–100
        estimates, 95–96
        evaluation by inspection, 97–98
        evaluation by integration, 98–100
        Riemann sums, 95

Green's theorem for, 87
per area, 90
Functions (elementary properties), 19–23
linear, 19, 23
of n variables, 22
of three variables, 22
of two variables, 19

Gauss, theorem of (divergence theorem), 102–107
Gradient, 29–36 (see also Directional derivative)
of a function of four variables, 36
of linear functions, 29–30
of non-linear functions, 31–36
used for estimates, 31, 33 (see also Linear approximation)
Green's theorem, 87–89
and the divergence theorem, 106–107

Helicoid, 15

Intersection:
cone and line, 17
curve and plane, 10
cylinder and line, 6
line and plane, 6, 11
of planes, 13
surface and plane, 20

Jacobian matrix and determinant, 122–133
of linear transformations, 121–122

Lagrange multipliers, 46–47 (see also Constrained extrema)
Level curves, surfaces, 19–23 (see Functions)
Line (straight line), 3
Line integral, 75–83
of scalar functions, 75–76
estimates, 75
evaluation, 76
Riemann sums, 75
of vector functions, 77–83 (see also Ch. 11, 14, 15)
curl-test in the plane, 81–83
estimates, 77
evaluation, 78–79 (see also Curl-test)
orientation convention, 77
path independence, 80
Riemann sums, 77
various notations, 79
Linear approximation, 34–37
function of four variables, 36
interpretation, 34
notation, 34
numerical computations, 36–37
Linear transformations, 117–122, 128–129
Linearization, 29–37
gradient and directional derivative, 29–34
linear approximation, 34–37

Normal component:
of the gradient, 88
of a vector field along a plane curve, 85
of a vector field on a surface, 93–95
Normal direction of an oriented plane curve, 85
Normal direction of a surface, 93

Objective function, 46
Optimization, 39–47
constrained extrema, 44–47
definitions, terminology, 39
function of one variable (review), 40
functions of three variables, 43–44
functions of two variables, 41–43
local extrema with zero discriminant, 43
Lagrange multipliers, 46–47
Oriented curve, 44, 77, 85

Parameter representation, 3
of curves in non-cartesian coordinates 10
of curves in the plane, 7–8, 10
of curves in space, 8–10
of lines in the plane, 3–5
of lines in space, 5–6
of planes, 11–13
of surfaces, 14–18

Parametric equations, 3
Parametrizations, different:
    of a curve in the plane, 8
    of a line in the plane, 4, 5
    of a line in space, 6
    of a plane, 12–13
    of a surface, 17
Partial derivatives, 25–27
    used for estimates, 26
Polar coordinates, 10, 49–51
    area element, 50
    double integral in, 60
    interpretation as mapping, 49, 122
    line integral over a curve in, 76
    linear approximation, 122–123
Position vector, 4

Speed, 5, 6, 8
Spherical coordinates, 10, 52–54, 128
    interpretation as mapping, 52
    line integral over a curve in, 76, 79
    triple integral in, 66
    volume element, 66
Stokes, theorem of, 109–113
    and Fundamental Theorem of Calculus, 113
Surface:
    area, 11, 14
    closed, 111
    integrals of scalar functions, 91–93
    parameter representation, 14–18
    of revolution, 16, 17, 42, 99

Torus, 102, 111, 113
Traveling wave, 20
Triple integrals, 63–66
    estimates, 63–64
    note on definition, 63
    Riemann sums, 63
    setting them up:
        in cartesian coordinates, 64–65
        in cylindrical coordinates, 66
        in spherical coordinates, 66

Vector, notation, 3
Vector fields:
    in cartesian coordinate system, 67–72
    conservative, 81–82
    flow lines, 69–70
    in non-cartesian coordinate systems, 72–73
    normal component:
        along a plane curve, 85
        on a surface, 93–95
    tangential component along a curve, 70–72
Velocity vector, 8

Notes

Notes

Notes

Notes

Notes

Notes

Notes

Notes

Notes